HELLHOUNDS OF FRANCE

This work is in the public domain Republished
by DVX Publishing Company, 2022
First Printing 2022
Cover art by Friendly Father
Paperback ISBN-13: 978-1-957583-07-5

HELL HOUNDS OF FRANCE

BY

EX-LEGIONNAIRE 1384

Of the Legion Entrangere

In. collaboration with

W. J. BLACKLEDGE

LONDON
SAMPSON LOW, MARSTON & CO., LTD.

Second Impression

MADE AND PROTTED IN GREAT BRITAIN BY PURNELL- AND SONS AULTON
(SOMERSET) AND LONDON

TABLE OF CONTENTS

To

THE MEN

WHO NEVER HAVE AN

ARMISTICE

CHAPTER I: AN INDICTMENT AGAINST FRANCE

This is an indictment against France. It is written in the centenary year of the French Foreign Legion. God alone knows what havoc among men this bloodthirsty body has wrought in the past one hundred years. No man may think of its crimson history without feelings of loathing and disgust against the nation that tolerates and perpetuates such a tradition of savagery.

As one who has been a Hell Hound in the pay of France, I accuse her of fostering and maintaining a fighting force the name of which, as she knows, must stink in the nostrils of decent men the world over. I accuse her of encouraging an organization which can only make inhuman brutes of men, which spreads disease, destruction and death wherever it moves.

Then, why did I join the Legion? I must answer that very natural question by asking another. Would any man join the Legion if he knew just what it meant? It is impossible to know the Legion without serving in it. It is equally impossible for any man to tell the truth about it, after serving in it. For there are many phases of that hell on earth which cannot be described in cold print. I can but give my own censored experiences.

I endured that living hell for five years—before I managed to escape. Three and a half of those years were spent in the Legion proper, and eighteen months in twenty-two French prisons. It is not my intention to give an account of the brutalities experienced in each of those prisons. Such a recital would be too monotonous. Nor shall I dilate at any length on the barrack life of the Legion, since that is by now fairly well known.

This account is largely concerned with the various engagements in which I took a part. It is concerned with the Legion in the field, in action, fighting battles and defending outposts on both the

Moroccan and Syrian Frontiers. Not all the engagements in which I participated will be dealt with, for this is no mere history of war. It is an attempt to show the life of the Legion in the field. For my sins, I was awarded the *Croix de Guerre,* the Medal of Morocco, the Medal of Syria, and the Rashaya Medal, which was specially struck for the survivors of that terrible blunder. I do not mention these decorations because I am proud of them. But it should be obvious that I must have been in the French Foreign Legion a few years in order to earn them. Such facts are recorded against me as Legionnaire No. 1384, and can always be verified.

I realize that much of what I am writing will be open to the severest criticism, that when I describe the purposeless bloodsheds and murders, the innumerable floggings, the mutilations and barbarities, immoralities and atrocities, for which the Legion is responsible, I shall be doubted. I shall be doubted when I tell that, on occasion, we were driven to slaughtering domestic animals for food, chewing bits of leather because we could not find water to quench a maddening thirst, hanging up live women for targets because they had mutilated our comrades as only Arabian women know how. . . . But the only doubt that should be expressed regarding a chronicle of this kind, is the doubt as to the sanity of a nation that fosters such an uncivilized force as the French Foreign Legion.

The men of this force are the most amazing crowd it is possible to meet. They are recruited from all parts of Europe, America and the Near East, from every class and every nation, and they comprise thieves, murderers, bandits, cut-throats, dukes, counts and scum. *En masse* they are the finest fighting force in the world. Individually they are the biggest scoundrels on earth. While you are standing up, they are the best comrades a man could possibly have. But as soon as you drop out, they will thieve the boots from your feet. They spend most of their time cursing the French and devising means of escape. But when they go into action they do so with the cry "*Vive la France!*" on their lips. *Soldat de la Legion Étranger! Vous n'avez pas de nation. La France est votre Mere!*[1] And, God help them! What a mother! What pride of family has she? Is she mad that she must gather the scum of the earth and smother it with a pseudo fighting patriotism? And to what end? The protection of a scrap of land in Africa, and a smaller scrap of Arabia! What would the civilized nations have to say if Britain adopted the same means of protection in Egypt, India, and Mesopotamia!

And yet, when an ex-legionnaire tells of things seen and experienced, things that happen every day in those French Protectorates (ironic title!), he must do so with extreme care, not so much because he may offend France or her Foreign Legion, but because, as my collaborator has repeatedly pointed out to me, there is such a thing as plausibility!

That is precisely the point! The Legion is beyond plausibility. It is beyond ordinary human conception. It stands out as a thing apart from any other sort of activity in life as we know it to-day.

All that I can do is to represent myself while being within the limits of the printed word. And in this respect, it should be appreciated that I did not join the Legion to study the problem of its existence, but merely to get a kick out of life. If, for instance, I have given a wrong impression of a particular tribe, of the Riffs, the Bedouins, or the Jebel Druses, I can but answer that it was not my job to inquire into their customs and morals, but simply the soldier's job and an unusual soldier at that! of facing up to them with a gun.

The name I use here is assumed, so is that of McCann, my American chum, Bess and Rehnke, the Germans, Lieutenant Beaumont, who committed suicide, and Captain Dupont, with whom I served in the Legion. For the rest, the names of men and places, of battles, massacres, sieges, and the like, are true, and, in the case of the latter, on record. Therefore, the whole of this account, save for the names mentioned, is true in substance and in fact.

My first engagement, which was at the outpost of Bab Mezab, on the Moroccan Frontier, was one of the most ghastly I was to experience, not because of the intensity of the battle that raged there, but because of the peculiarity of the conditions. For the first time in my life, I came to know what it meant to exist in the burning heat without water for days on end, *and fight!*

I propose to plunge the reader straight into the middle of that terrible affair, which was my Moroccan Baptism.

CHAPTER II: MY MOROCCAN BAPTISM

I can never recall that battle of Bab Mezab without picturing my American chum, McCann, lying sprawled across the broken wall of the fort, mouthing the foulest oaths imaginable.

He cursed, not like a trooper, but like one demented, as he adjusted the sights of his rifle; and though his eyes were wild and bloodshot, his great hands were perfectly steady. He was enjoining me to pray. His favorite expression when addressing me was "You blighted limejuicer!" And during that particular incident he was using it with sickening repetition.

"Pray, you blighted limejuicer! Pray that I don't miss those two poor devils out there! D'ja hear, Barrington? Two shots ... no more . . . one for each . . . and away they go . . . out of their cursed misery!"

I answered that I *was* praying. In those tense moments his face was incredibly ugly. The ferocious light in his eyes, the scarred, unshaven jowl, the sweat that had turned dust and sand into tiny riverets of mud on his cheeks, the partially open mouth trickling saliva this could not be my American pal of the French Foreign Legion. It was some indecent brute grown out of the desert wastes.

I watched him take careful aim at one of these two ghastly targets, grotesquely stuck there, in the sand, some five hundred yards beyond the fort—this fort of Bab Mezab, which we had held so desperately and so hopelessly for six days and nights.

Behind us crouched Captain Dupont, commander of the Fifth Squadron of the Legion cavalry, or what was left of it. He had chosen McCann for this devilish job because he knew that McCann could not miss at five hundred yards. McCann was a crack shot. It was imperative that he should register a clean hit at each of those objects, that there should be no agonizing wounds further to torture the pitiful creatures. A clean hit to the brain, quick, merciful, instantaneous death for those two screaming objects that had, less than an hour ago, been our comrades, fighting alongside us in the fort of Bab Mezab.

Bab Mezab! Was there ever such an insane blunder? We of the Fifth Squadron, beginning with 250 men, banded together to hold this fort against unnumbered brutes from the Riff mountains, 250 wretches of the French Foreign Legion, comprising every nationality under the sun, we were fit only for the dirty work of the French Protectorate of Morocco. That, I found, was the legionnaire's first lesson.

Well, we had held the fort so far, six days and six nights, fighting, always fighting that yelling mob of Riffs down there, on the plains. Abd-el-Krim, powerful chieftain, had declared war on French and Spanish Morocco. He had mobilized his hordes of madmen and fanatics, soldiers of the Riff mountains who had been trained and disciplined by German adventurers, and armed by cannon of British manufacture, and had descended from the mountains to the chain of forts, of which Bab Mezab was one.

We had passed from one outpost to another of the series that run down to Spanish Morocco. We had force-marched by way of Tunis to Mequinez, then to Fez and Tasa, taking in outposts on the chain, rushing from one to the other like halfwits, until a flying column under General de Shambrum had rushed us to the fort of Bab Mezab, and left us to our fate, to hold the fort at any cost.

And at what a cost! Day after day, night after night, the vast army of Riffs fired shot and shell into the little fort. The telephone communications were cut. We were trapped in our stronghold. The Riffs knew, and we knew, it was but a question of time. We had lost more than half our men, wounded lay writhing in agony, for the Legion carries no ambulance corps, no organized medical section such as may be found in any other fighting force. Every legionnaire is his own M.O. For the rest, there are N.C.O.'s with a crude knowledge of medicine and an even cruder knowledge of surgery.

It was not the fighting, however, that worried the Legion. The Legion was there to fight. The Legion loves to fight. The Legion is the finest fighting force in the world. The rations had run out. That did not worry us either. But it was the temperature of 106 in the shade. The water had run out. And that did worry us! Like Gehenna, it did. We sprawled the crumbling walls of the fort of Bab Mezab like men bemused, lips cracked, tongues lolling, mouths gritty, pressing the clips, firing, pressing the clips, firing . . .

A couple of hundred yards from the walls of the fort was a waterhole. Water there! And we had lost eight men who had tried to reach it. Captain Dupont had lost too many men. He would not risk any more. Men, crazy with thirst, volunteered, begged to be allowed to run the risk of capture and torture by the Riffs, in the hope of making that waterhole, filling a dixie, and getting back within the fort. Captain Dupont's answer was typical of the man. He lifted his boot to the belly of the volunteer. And that was that. Even though his own lips were cracked, his own mouth scarred with grit, he would not risk another man. He himself preferred the intolerable thirst.

There had been a waterhole inside the fort when the Fifth Squadron took over, but the Riff gunners had found the exact position of it. Those British-made guns, and the German officers behind them, had done the rest. That waterhole became a mass of broken masonry and sandy mud. The only water left now was that which lay some two hundred yards beyond the fort.

Then it was that we had seen the ghastly spectacle which brought into prominence the expert marksmanship of my pal, McCann. Down there, on the desert plains, some three hundred yards beyond the waterhole, was a band of Riffs with two of our comrades who had been captured while trying to get the water. In the sun's glare we saw every detail of that damnable torture. We saw the two legionnaires, bound hand and foot. We watched in sickly horror the digging of two holes, deep and of small diameter, so that a man could be stood up in each, the earth filled in, and only the head of each man showing above the face of the desert.

We watched the Riffs approach those two heads sticking up above ground. We saw them pour over the two heads a brown sticky substance, which soon we knew to be honey, for in a moment the two heads were covered with a cloud of flies, mosquitoes, scorpions and other torturous insects. We heard the inhuman screams that came from those heads, screams that could not be drowned even by the wild yells of the Riffs, whose insane sense of the comic had prompted this diabolical torture.

And from the walls of the fort there had arisen other yells, yells of maddened white men. Captain Dupont had rushed among his cursing men, saber in hand, threatening to behead any man who tried to rush from the fort to the help of those tortured heads. Then he had given the order to McCann, crack shot of the Squadron, to put an end to those two legionnaires who had tried to get water for the besieged fort. And McCann, six feet of lean muscle and grit, an adventurer who lived only to fight and drink, who should have been occupying a desk in his father's office away in Detroit, McCann, the lovable scoundrel, had taken aim in the coolest manner imaginable, even while he mouthed foul oaths and enjoined me to pray that he should not miss his targets.

"Pray! You son of a gun!" he said. "Pray that I don't miss!"

And I did. His rifle cracked. A shower of insects shot up into the air. A head lolled forward, blood streaming from the eyes. Captain Dupont grunted. Crack! McCann's rifle spat again. The other head shuddered. Another shower of flies. The insects descended, but the heads on which they fed could feel no more. The enraged Riffs advanced, yelling their battle cry.

"Come on, you!" cried McCann. "Come on, you swine! Torture legionnaires, would you!"

And come on, they did in amazing numbers, to be mown down like rotten sheep. They could not be checked. They used their dead as shields, and fought like maniacs. They dropped in the sand like so many flies, only to be reinforced from their rear. If only we had water, I thought, we might stick this for another six days and nights, spraying the desert with dead Riffs. If only we could get a drop of water, just a drop, to keep us on our feet.

There were so few of the boys on their feet now. We had lost four out of seven officers. There were fewer rifles at work now. I could only guess all this by the sounds, or lack of them. A legionnaire never turns round, never looks back, while in action. If he does, he does it only once. Captain Dupont has a quick eye for that type of man. He could draw a sawn-off Mauser quicker than any man; and the legionnaire who turned his head, never turned again.

The Riffs were advancing, inch by inch. The desert was strewn with their dead, piled high here and there, like so many fortifications. A dead Riff is just so much lumber, to be used if necessary, as lumber. It is his soul that matters. And that is safe in Allah's keeping. Every Riff wants to die fighting, fighting the infidels. To him, there is no more beautiful death, no surer way to the throne of Allah.

Captain Dupont was giving orders, addressing each legionnaire individually, prefixing his commands with his customary epithets: “Barrington, *nom du chien!* Do this, do that! McCann!" who was swine, dog and cow, all rolled into one, "Yelia!” The Skipper’s eyes were wild, malignant. His face was streaming with sweat and blood. But he was in his element. He was quite as mad as the Riffs. What other sort of Skipper would have held Bab Mezab as he did? He put his face on the wall beside me. For a second, I thought he had stopped one, but only for a second. His harsh, grating voice was bellowing in my ear, trying to make itself heard above the din and roar of the fighting devils down below us. The order was to move! We were running now, from one embrasure in the fort wall to another, stopping only to fire, run, stop, fire! That was the order. Run, stop, fire. A wretched handful of men were racing round the inside of the fort, trying to fire from a dozen places at once, miserably trying to make the hordes of Riffs down there believe we were a vast column, instead of a score of thirst-maddened, half-demented legionnaires.

The ruse had its effect. At all events, the Riffs, who had begun to ascend the hill towards the gates of the fort, fell back. They retreated beyond range. We ceased fire and watched. Bands of them were coming cautiously forward. They began to collect their dead. We let them. We had no ammunition to throw away on targets already accounted for. Besides, another hour would see the great ball of fire dip into its bath of sand, darkness would descend with the suddenness of a dropped curtain. And who knows what this night would bring?

Four men were told off to patrol the fort wall. The rest of us sat down on the ground, exhausted, indifferent, staring at the bloody mess within the walls of fort Bab Mezab. Dead and dying lay sprawled, spread-eagled, crumpled, so many ghastly shapes. Wounded men were screaming for water. *"L'eau! L'eau! L'eau !"* But no man stirred, nobody moved, except the four on guard. They continued their patrol, back and forth, back and forth, mechanically, like so many robots. What was the use? It was only a question of time, and the score of us who were left would join them. It was easy to feel indifferent to the yells of pain, easy to ignore the cries for water, for thirst-maddened men are dead to all human appeal. What did it matter that men died with foul oaths on their lips, or that live men sat sullenly watching?

Suddenly one of the boys who had been resting with us, a Polish boy hardly out of his teens, named Tomski, sprang to his feet, and leapt upon a wounded legionnaire whose screams were too incessant, beating him with the butt of his rifle. The screams died suddenly. A shot rang out. Tomski slued round. His ear was dripping blood. He was stupidly trying to put it back on his head. He was staring at the Skipper's Mauser, and the whites of his eyes told their own tale.

He had gone *cafard* we were suddenly roused from our sullen indifference. Desert madness had to be dealt with. It brooked no delay. Poor little Tomski! He had seen more at nineteen than most men see at ninety. Now he saw red. He ran amok. His rifle spat four times in quick succession. Four more legionnaires went home before he was caught, and the Skipper's Mauser had put an end to his maniacal outburst. It was all over in a few minutes, but he had done more damage to the Squadron than the Riffs could do in as many hours.

We dropped back on the ground. The guards returned to their posts. Sixteen men, waiting for darkness, and the end. I looked at Lieut. Sera, the only officer we had left, save the Skipper. He was whimpering noiselessly. Would he be the next to go *cafard?* Better keep my eye on him. He, too, was young, almost as young as

Tomski. But, unlike Tomski, he had joined the Legion to fight for France. Tomski had joined the Legion to see the world. He was built that way. I could recall Tomski's boyish, wide-mouthed grin, when asked why he joined the Legion.

To see the world! Well, he had seen it. And now he lay over there, in a pool of blood. No more looting native villages. No more hunting of native girls. No more smutty stories. Tomski's mouth was closed for ever. He was done. Packed up. Finished. Nineteen eventful years, and then phut! All over. And what years—if he were to be credited with any sense of veracity at all! years of passion that began almost at school age, two passions, in fact, one for fighting and the other for women, and beyond these, no horizon at all. Poor Tomski!

High above us the vultures were hovering. They, too, knew the end was near. They were only waiting to feed.

"Barrington?"

I turned and gazed at McCann.

"Listen, you limejuicer," he said, "you're the Skipper's servant, and you've always got a gun."[1]

"Well, what are you getting at?" "Listen, boy, if I stop one, give us your word you'll put me right out. You know what it means if you're caught alive by those devils down there."

"All right, McCann, boy. I promise."

We looked at each other for a few quiet moments, a far-seeing of look, filled with meaning. We had seen others tortured. Better sort dead than alive in the hands of the Riffs. Honey and scorpions was by no means the worst kind of torture. We had buried men in the fort three days ago, who had been mutilated by the Riffs, and then dragged back through the sand, in the dead of night, almost up to the very gates of the fort. Certainly near enough for us to collect them, and gaze upon the handiwork of

[1] An officer's servant in the Legion is also the officer's bodyguard, and is always armed.

the Riffs. Three headless men tied together by their ankles. Headless but that was not all....

"And I'll do the same for you, you old son of a gun."

"All right, you blighter. Shut up!"

"Yeah? Shoor! What's biting you, Barrington? Don't you like this outfit?"

"Well, of course! I'm only sorry we can't stay."

McCann croaked through his cracked lips. It was the best he could do in the circumstances.

"Oh, boy, what wouldn't I give for a bidon of hooch."

"Every bidon was emptied two days ago; and anyway, you big stiff, you've had enough wine to last any ordinary man a lifetime."

"Yeah?"

"Yes. Even if you conk out tonight."

"Be cheerful, brother, be cheerful."

McCann's India-rubber face was twisted in a wide grin.

"We're going to get out of this, boy."

"How?"

"How!" snorted McCann. "How! How, how, how! Just like you, you goldarned Britisher! Bark! That's all, bark! D'ya think I'm going to let any mother's son of a Riff get me, eh? "

"I admire your pride of nationality, McCann; but it'll take more than one American to stop that mob down there." "Yeah?"

We did not continue the futile argument. The Skipper's boot was too active. He wanted four "sons of dogs" to relieve the guard, and for two of them he chose the English "cow" and the American "swine." It was sometimes hard to believe that Captain Dupont loved us like a brother. But he did, in spite of his tongue and his boot. McCann and I chose posts where we should be able to meet every twenty paces. We were the only two English-speaking legionnaires in the Squadron, and our companionship was taken for granted. Marching up and down, we grew quite cheerful, almost human, getting some slight relief from the torturous thirst by chewing pieces of leather. Up and down, up and down, every twenty paces or so I met McCann, face to face, he would sum up my British nationality in a few choice words, and about turn, to pace up and down.

There was little to attract one's attention, save the cries of the sun-scorched wounded, which were getting fewer and fewer, and the vultures sailing overhead. It was late afternoon and the sun was pouring forth its greatest heat before disappearing. Down in the desert there was a strange, sinister lull, hardly a movement. The naked eye could barely discern the moving shapes in the distance. What were they preparing for, down there? A night attack. That was certain. But at least we should be free from the glare and heat of that fiery ball.

It sank suddenly, like a weight in water. Water! God! If only we could get out to that waterhole under cover of this gathering darkness. One gulp! Just one! And how happily we should fight them! I put it to McCann.

"Ask the Skipper," he grinned, knowing what that meant.

The fort was now a mass of shadows, some of them sprawled and motionless, others moving about silently. The sizzling heat was gone. That was a relief. Every man was on guard now, fourteen men, two corporals, two officers. They were watching, waiting. Black masses were moving down there, leisurely, it seemed, as if the murder of sixteen legionnaires, and two French officers were a matter to be conned over, deliberated with anticipatory pleasure, carried out with a subtlety and finesse appropriate to the Riff mentality.

Captain Dupont was leaning on the wall, trying to focus his glasses on the dim figures moving in the sand. One of our men approached him, a big, hulking, black-bearded brute, familiarly known as the "Count"; one of those hundreds of fellows to be found in the

Legion who are hereditary monarchs in their own rights, if only they were permitted to assume their rights. The Count was a Bulgarian, a fearless fellow, hard fighter, deep drinker, wonderfully adept with the use of a bayonet, whether it was opening a tin of beef or a Riff carcass.

McCann nudged me.

“That guy’s after the waterhole,” he grinned.

I nodded. Anything for a diversion.

We watched the Count approach the Skipper, heard something of his ragged muttering. The Skipper slued round and brought his field glasses down on the Count’s head with a crash felling him like an ox. He lay there, his tongue lolling near a pool of blood that had trickled from a dead comrade.

“Hell! Look at that! ” Said McCann. “ Don’t tell me the son of a bitch is going to lick it!”

A corporal saw it and, emulating his master, began to kick the Count until he had induced him to get on his feet again. “Say,” murmured McCann, confidentially, “it’s a bit of a descent for the Count, this, what d'ya call it—vampirism.”

“Oh, for God’s sake, Mac! *You* filthy blighter!”

“Steady, boy!” with intense surprise, “what’s on your mind?” And then the fun began. The Skipper’s blow and the corporal’s kicks had proved the last straw for the Count. He went *cafard.* He had wrenched out his bayonet and had put paid to the corporal’s ticket before any of us had time to realize what was happening. Captain Dupont let out a bellowing roar and fired. The Count ducked, and stuck another of his comrades in the process. We daren’t close in on him. The Skipper’s Mauser was too active. In the main, we were dancing about helplessly in the dim light. The Count was using his bayonet like a man possessed, and to some effect, for there were four dead men at his feet, in addition to the corporal, before McCann, risking the flying shots, had cracked his skull from behind.

The sound of the shots seemed to be taken as a challenge by the Riffs. They opened fire on the fort, and Captain Dupont was booting every man to his post. From now on, I thought, there would not be time for any more legionnaires to go *cafard.* We should be too busy. There's an old biblical tale, something about the devil finding work for idle hands. And doesn't the devil get busy among legionnaires who have nothing to do except nurse a perishing thirst! If only we had a drop of water! If it had not been for this problem of water we should still have been a pretty tough bunch to reckon with, instead of which we were starting this unforgettable night battle with ten men, one N.C.O., and two officers, thirteen all told. . . . I must have been crazy then.

"Mac!" I shouted, "we're thirteen men standing!"

But McCann was in full war cry. He was yelling at the Riffs. "Come on, you ..." Not that they needed any invitation. Up they came. These sons of Allah were fighting for their faith. They had long prepared for this, and were determined to rid Morocco of its infidels. What, after all, was Bab Mezab? Just one mean little fort. If reinforcements came now they would be too late. They would be mown down before they could reach us. Strange how one clings to the merest thread of hope. . . .

Bullets were stabbing the darkness in all directions. The Riffs were firing at random. Since they could not see their targets, they were out to impress us by the volume of their fire. We, on the other hand, were sparing with our fire, moving round from one position to another, like the ghosts of a lost army, sniping with effect, rather than fighting.

And we realized that we were holding them. After the first mad rush up the hillside, accompanied by screaming shells that had done nothing worse than burst the masonry of an empty magazine room, they had dug themselves in the sand behind their dead. Their big guns were silent now, obviously afraid to short-range on their advancing column in the lowering darkness.

Hour after hour through that ghastly night they kept up the blind attack, guessing, no doubt, that our numbers had sadly dwindled; but even they could not know that we were but a baker's dozen, or they would not have hesitated. There would have been one savage charge, and the Fifth Squadron of the Legion Cavalry would have come to an end.

Perhaps. We had our trump card in the shape of four machine guns, all ready to spray death when, in the last advance, the Riffs came near enough to receive it. All night long the Riffs kept up an incessant fire, and we a perpetual sniping. Wearing us out, of course. At dawn would come the final stroke. How we watched for the first streaks of morning light! "Another half-hour, Barrington, and we'll see daylight," said McCann, in a hoarse and weary voice.

"Yes, in more senses than one."

"Cheerful cuss!"

That first streak came, a thin penciling of light. We crouched over the machine guns, waiting. Suddenly the heavens opened in a brilliant splash of color, orange, crimson and violet. As we had anticipated, it was the signal for the enemy to make their last mad rush on the fort. Up they came, a solid black mass, screaming and yelling like maniacs. And our guns replied: Rat-tat-tat-tat . . . tattat . . . tat . . . tat . . . tat-tat-tat-tat-tat! Such music, to weary and exhausted legionnaires.

The vultures were early astir, sailing low over the hillside. And then, out of the blue, came another and different vulture, dropping death in its wake. A French bombing 'plane! Then another, and another! They were shattering and scattering the black mass beneath them. Riffs were tumbling down the slopes in a mad race with death. Their column was thrown into the wildest confusion by this quarter of death, guns were wrecked, and as they retreated they were met by our own relief column, which literally mowed its way through to the gates of Bab Mezab. At the head of the column was a battalion of Legion infantry. They had come to take over. God help them!

They found us a dozen starved, thirst-maddened wretches, sprawling over the waterhole that our hungry eyes had gazed upon for days, or was it weeks? Men of the Legion infantry crowded round us, took off their bidons full of good red wine and handed them round. We drank and drank and drank. McCann emptied three in quick succession.

His eye sparkled as he turned to me. "You godarned lime juicer, Barrington! Didn't I tell you we'd get out of this ten cent racket?"

CHAPTER III: A FRENCH OFFICER WANDERS

We left the "ten cent racket" as McCann termed it, within a few hours of being relieved. There was no time for sleep, or even rest, and hardly time to de-louse or wash off the six days' accumulation of grime. We should have time enough for all that in Fez, we were told, and it was essential that we should get there with all possible speed.

We realized how imperative was this forced march as soon as we started out, for our escorting column was a bunch of Senegalese and a company of infantry. We also had a couple of airplanes to keep the sniping Riffs from coming too close. We took the center of the column, on foot like the rest, for we had lost all our mounts early in the Bab Mezab blunder.

Imagine us, ten wretched legionnaires whose clothing was almost in rags, and two officers in no better plight, a dozen men who had scarcely slept for a week, tramping along in the middle of that column, some with dysentery, others with malaria, and nearly all with minor wounds, everyone a sick man, for we had gone hungry and thirsty for three days, and had then filled up with water, beef, biscuits, and far too much wine from our comrades of the Legion infantry who had relieved us.

It may seem incredible that such a band of men as we were then, underfed, sick, diseased, exhausted, could face a day's march after all we had endured, but life in the Legion sets at naught all accepted laws of human endurance. The explanation of this is perfectly simple. While the column is on the march no legionnaire would dream of falling out, unless, of course, he were *cafard,* and therefore unaware of the meaning of a lingering, torturous death.

Usually a slash from an N.C.O.'s whip is enough to waken a laggard. If, after that, a man falls in the sand with the intention of remaining there, if, despite all his efforts, he is unable to rise, or if he is attempting desertion, he is invariably found by the Arabs or, perchance, the Partisans. If found by Arabs or Riffs, he wall never be seen again. The Partisans will convey him back to headquarters in a fashion. And what a fashion!

These Partisans, "Headhunters," as they are commonly known in the Legion, are parties of Arabs recruited for the purpose of policing the desert. They are paid one hundred francs a head for every man they bring back to the fold. God help the sick man, or the deserter, who falls into their hands. He is always returned to headquarters in a frightfully mutilated condition, and the Headhunters explain, with a characteristic shrug, that such was the prisoner's condition when they found him. The truth of the matter is, of course, that they themselves have committed the atrocities for their own amusement. Having satiated themselves in this manner, they hand the unfortunate devil over to the authorities, and he either spends a few weeks in hospital, or dies, a much more likely event, or a few years in chains in prison. It is a matter of some moment that, however many years in prison chains a legionnaire may do, the time is never reckoned as service with the Legion. That is to say, if a man has done two years in the Legion, and is then sentenced to four years in chains in prison, he has still to serve three years in the Legion before completing the five for which he was enlisted.

Is it any wonder, then, that Legionnaires rarely fall out, however sick they may be, however harsh the conditions of the march? Is it any wonder that men drop dead on the march? During my service with the Legion I have seen scores of men die on their feet. It is a ghastly experience to be marching alongside a man severely wounded, or racked by dysentery or enteric, to be helping that man to keep on the march, and suddenly find his body grow heavy, the mask of death in his face. Yet that happened to me far more times than I care to remember.

To march with a temperature that would put most men on the danger list, is a common experience, for there is no danger as long as a man keeps moving with the column; the danger lies in falling behind and staying there, since that means something worse than death.

There was, however, no falling out among the ten survivors of Bab Mezab. We slouched along through the intolerable heat, in the center of the column, and in the middle of a column of dust, too. All day long we marched, the Riffs sniping from nullahs and sand-drifts that even the 'planes could not locate. How we prayed for nightfall which would mean an encampment of some sort, and blessed sleep.

McCann, stumbling along by my side, became strangely quiet. He was asleep, or at least, dozing. He was one of those amazing people whose limbs may move mechanically while the brain sleeps. On my other flank a young French officer had fallen into step. He marched with me in silence for a while, and I wondered not a little at his maneuver, though I was too sick and weary to trouble too much.

Then this young officer began to talk. He was Lieut. Beaumont, and what did I think of the Moroccan war? Naturally I thought he was merely amusing himself with a fool of a legionnaire, and I took little notice of his chatter, did not trouble to answer, in fact. But I gathered that he was fresh from St. Cyr, the Sandhurst of France, and was attached to the battalion of Senegalese. His appearance at Bab Mezab with the relief column was his first taste of the war zone.

"Mon *Legionnaire, c'est tris terrible! Terrible!"*

I stared at him then. He was only a boy. He was nervous, excitable. Apparently the experience had been too much for him. He felt he had to talk to someone about what he had seen up there in the fort, and in his state he could hardly talk to a brother officer, not if he wished to keep his self-respect. How on earth did he get into Morocco, I wondered. He should not have been allowed to leave school for another few years.

He said: "You look as if you had been through hell."

"Not hell, mon Lieutenant, but something worse!"

"Do the Riffs always cut up the wounded like that?"

"Yes," I admitted, "but sometimes the tortures are more horrible."

"But they could not be more horrible!"

How little he knew, this frightened little son of France. He was chewing his lips, making vain efforts to keep his nerve. Even in my utterly weary and fed-up condition, I could not help feeling a twinge of sympathy for the kid. God in heaven! What would happen to him, if he were caught by the Riffs!

He did not address me again during that soul-searing march but just plodded on, muttering and mumbling to himself. I thought if this kid goes *cafard,* I'll brain him at the first opportunity. But he did not. He kept on marching and muttering, marching and muttering . . .

Darkness fell and we camped down for the night. The little band of legionnaires kept together. We were given a double ration of wine and bread. Some of the French infantry, despite the fact that they hated and feared us, came and fraternized, something they would not dream of doing in bar racks or elsewhere. They gave us more wine, and passed round cigarettes. For one night, at least, the despised Legion was to be respected. It was indeed a novel experience. The wine flowed. It did not take McCann long to get tight. That man must have had a constitution of cast-iron. He began to sing songs about Dixie and his Coal Black Mammy, and the French men stood about and listened, believing him mad. Perhaps he was, a little.

In the middle of the carousal I fell asleep, only to be awakened somewhere about midnight. There was a terrific commotion in the camp. I grabbed my rifle and jumped to my feet. McCann was already there.

"O.K., Barrington" he grinned, "there isn't a Riff in sight. Kip down again, you lime-juicer!"

"What's on?"

"Some guy's dropped his medal."

"Who?"

"That young Lieutenant pal o' yours, ain't he? Oh, boy!

"He's bolted!"

"Lieutenant Beaumont, d'you mean?"

"Shoor! And I don't blame the fool. Nice little boy. But it'll be a bit tough on his folks, eh? Pass the bidon, Barrington, and for the love of Mike, don't look so damned scared!" "But, McCann, he's only a child!"

McCann grinned. "That's what the Riffs will take him for." "Mac, you're a dirty swine!"

"Yeah? Waal, I guess there's a war on, round these parts."

"I can't get that kid out of my head."

"Say, boy, take a drink and kip down again. We shall be hoofing it in another hour or two."

I rolled into my blanket, but in spite of my numbed and aching limbs, I could not sleep. I was thinking. I stared up at the stars with the picture of that French boy in my mind, wandering about in the desert alone, a prey to unnamable things. Why should I worry about the French kid? Why did he want to come out to Morocco for, when he might have been doing garrison duty in the south of France? Men of the Legion have definite reasons for their presence in Morocco or Syria, mostly unsavory ones. So have the French officers who command them. But this kid . . .

When we assembled next morning to march on to Fez, word came that the missing Lieutenant had been found. Well, it was a case of God help him now in either event, lost or found. Even the legionnaires would despise him.

By noon of that day we had reached Fez, and were soon settled in the barracks. A large detachment of Legion cavalry awaited our arrival. The Fifth Squadron was to be reinforced without delay. Obviously there was more dirty work ahead for us. It would be interesting to get among a fresh bunch of legionnaires. There would be another batch of counts, princes, rogues and vagabonds, more hereditary monarchs fleeing their countries because of some vague, inexplicable crisis that seem always to be creating disruption among the smaller European states. There is this to be said of life in the Legion, it is never monotonous!

On the second day after our arrival in Fez there began that series of parades and inspections and preparations that mark the life of the Legion in barracks. The newcomers gained the idea, of course, that we were about to move off almost immediately. They were later to learn that such hurried bustlings and sudden alarms were a regular feature of the life, for the men of the Legion are never allowed to rest many minutes at a time, except in the evening, and most of those are taken up with patrol duty in the town.

Floggings, too, are a constant source of diversion. Freshers had to learn, in the most painful fashion, that a Legion N.C.O. must be saluted, whether in or out of barracks, and that they must stand to attention while addressing non-commissioned officers. Failure to comply meant a flogging. And the French are supreme artists in the use of the whip and the salt-water pail. They flogged scientifically. They flayed with the salted whip in such a way as to render a man insensible by means of a given number of strokes, since no flogging was completed until the victim lost consciousness.

Some of the sergeants took a great pride in this work. In the first place, it was invariably carried out under the eyes of the general body. Perhaps, as a sort of example to others. Secondly, the sergeant wielding the whip would usually keep up a stream of sardonic and ironic comments:

"Ah, mon legionnaire . . . slash! Did not understand, eh? Slash! Poor fellow! Slash! I am so sorry for you . . . slash! If I could but help you . . . slash! My heart aches for you . . . slash! Shall we tell momma? Slash!" And so on, until the body of the victim slumped into unconsciousness.

It was on that second day that McCann and I were sent for by Captain Dupont. We had been chosen for special guard over an officer in prison. We were to do the unusual guard of three hours on and two off, and the Skipper gave us to understand that we had been picked because we could be trusted to keep our mouths shut.

Off we marched to the officers' quarters. Our prisoner was Lieut. Beaumont. I had suspected as much, and so had McCann. We had instructions not to leave the door of his room, and to spend as much time inside with him as possible. It was late afternoon and as I was mounting first guard, McCann decided to roll into his blankets and get some sleep. At the end of two hours he would take an hour with me, then two alone, and so on.

My first hour with the boy was spent in his room. He wanted to talk, and I made no attempt to check him. He was terrified, appalled at the thing he had done. He had lost his head, and had wandered off into the desert. Even now, he did not understand why he should be awaiting trial by *Conseil de Guerre,* since he had not known what he was doing. He had not been himself, that night, when he had walked blindly from the camp into the desert. A little mad, perhaps. Did they not excuse madness in the French Army? Surely he could not be held responsible for such an action?

"Mon Dieu! I can never stand the disgrace!"

"You will get over it, *mon lieutenant."*

"Non, non! Non, non! The dishonor! I shall be drummed out, or sent to that terrible penal battalion!"

He wept, talking of his beautiful home in Brittany. He had never wished to serve in the French Army, much less in Syria. But since he must serve, his father had insisted that he should see something of the world during the period. He talked of his mother, brother, and his two sisters. What would they think of him, a dishonored soldier of France? He would never see them again, never enter the home he loved, never enjoy again that wonderful garden in which he had played since childhood. He was ruined! Finished!

"Is it not strange that I, an officer of the French Army, should talk to you, a soldier of the Legion, like this? But, maybe, *mon legionnaire,* you, too, were once an officer? Why did you join the terrible Legion?"

I smiled. I wondered just how many times I had been asked that question! And because I had always refused to answer it, it was assumed that I had committed the worst of crimes, for the legionnaire who has no tale to tell of his presence in the Legion *must* have committed the capital crime, or why should he be afraid to discuss the reason?

But this young officer, with his sentimental turn of mind had quite another reason.

"Une femme?"

I laughed. A woman indeed! All sorts of reasons prompt all manner of men to leave their homes and their countries, to lose their identity in the Foreign Legion. But a woman? Well, indirectly, no doubt, there is often a woman behind these strange disappearances.

But this young, pitiful specimen of a soldier was not laughing. He was blubbering openly now, his nerves in shreds, blabbing about the little girl he would never see again, she of the so wonderful lips and eyes and hair. The fool! I was suddenly overwhelmed with disgust at the spectacle. So, it was a girl! He had been thinking of her, worrying about her, was tormented by the thoughts of her, when he had walked stupidly into the desert. It was not so much the mother, the brother and sisters, the beautiful home and the lovely garden. It was some chit of a girl. My godfathers!

I left him then. He would, he said, busy himself writing letters. He would write to his mother, and to the little girl. I took up my stand outside the door. Occasionally I looked through the grill. He wrote steadily, for nearly an hour. He wrote swiftly, as one in dead earnest, writing against time. There was quite a neat pile of letters on the bench. Even then I did not dream of what all this meant. The next time I looked through the little grill he was on his knees, praying. I turned my back on him. Poor devil! The next minute a shot rang out. I could not have been more startled had I been the target. Swinging round, I stared through the grill. Lieut. Beaumont was crumpled up on the floor, a smoking automatic in his hand, blood trickling from his temple.

I burst through the door and dropped down to my knees. He was almost gone. Nothing could save him now. McCann came flying into the room.

"Barrington! What in hell . . . ?"

An excited group of officers came crowding in, and with them, Captain Dupont.

"Sucre bon Dieu! Barrington, how did this happen? Who gave him the Mauser? *Nom du chien,* speak, speak!"

I explained that I had left him alone in the room, that I had looked through the grill and had seen him at prayer and then the shot. I knew nothing about the Mauser. I did not know he had one. We had had no orders to search him. We assumed that such precautions had been taken before we mounted guard. The Skipper knew, of course, that it was not our job to search an officer.

Captain Dupont listened, nodded, turned the body over with his foot, shrugged his shoulders, then left the room with the bunch of officers tailing him. McCann and I were dismissed. There was no further need for guard. That boy had taken the gentleman's way out. From the point of view of his family, he had decided it were better to die than live to dishonor them.

"I wonder who slipped him the gun, McCann?"

McCann's eyes lifted. "A sympathetic brother officer, I guess."

And that was that.

Early next morning an inquiry was held. I, naturally, was the principal witness. I could only repeat my story. Had the prisoner talked? He had talked a great deal. What about? I recited as much as I remembered. Captain Dupont showed his disgust. He asked a great many searching questions. Every officer in his battalion was closely questioned. The result was nil. Nobody knew, nobody cared. It did not matter, anyway. The inquiry was a mere formality. The obliging officer, whoever he was, had saved headquarters a lot of trouble.

The episode was dismissed with characteristic shrugs. The French officer has much in common with the Arab, the Riff, the Jebel Druses, and the rest of the tribes. He is a fatalist. He adopts the only sensible philosophy in the land of mystery, murder and sudden death. So also does the legionnaire. What other philosophy of life could be maintained for ten minutes in such an environment?

But it was not the end so far as McCann and I were concerned. We had hardly got back into barracks that same morning before we were hauled out again by a sergeant, and taken once more before Captain Dupont. For our sins we were detailed to accompany a couple of legionnaires out beyond the town to the open desert, where they were to be punished for some misdemeanor or other.

McCann was cursing volubly as we saddled our mounts and rode round to the prison quarters.

"What do these damned frogs think we are, Barrington?"

I laughed. "I expect they take us for a couple of legionnaires!" "Or a couple of first class jailers. I wonder what these two poor devils have been doing?"

As soon as the prisoners were brought out, chained wrist and ankle, we had a pretty fair idea. They were two Germans, hefty, tow-haired giants, as tough as you like, and ever lastingly rowing with somebody. I should think those two fellows had enough wheels on their broad backs to supply a railway track. But no amount of flogging seemed to affect them. They would turn up a few hours after leaving the whipping post, contrive to become fighting drunk, and start more trouble. They were good soldiers, and I suspect the Skipper was not at all inclined to send them to prison for any length of time. He wanted to keep them. But he was determined to cure them. As the British sergeant-majors used to say "You broke your mother's heart, but you won't break mine!" That was the Skipper's complex, but intensified a thousand times. His object was to break the two Jerrys, Behman and Sargas, and have done with their tantrums once and for all. Hence this little expedition into the desert.

We set off at a trot, the Skipper, a Russian sergeant named Lendedoff, a Belgian corporal named Peters, McCann and myself. We were all mounted, of course, leading the prisoners at a jog-trot, padding the sand with their chains jingling. Once we were clear of the town, the order was given to gallop, and what this meant for the two poor devils on foot must be seen to be fully appreciated. In less than half-an-hour they were drenched with sweat, coated with sand and dust kicked up by the mounts, and reduced to a state of utter wretchedness.

But we realized that this was only a preliminary to the actual punishment. We knew only too well what it was to be. We knew also that we should be in the desert for eight or nine hours, and we certainly did not relish the prospect. It was going to be a devilish long day. The inquiry had been held at five a.m., and it was barely seven o'clock now. The sun was beginning to make itself felt. We of the escort were fully protected against the fierce heat rays by our cap-curtains and, of course, full-dress uniform. (The legionnaire always wears a thick blue-serge tunic, despite the heat. It makes him perspire freely and is decidedly uncomfortable; but it stops the sun from penetrating to his spine.) Our two prisoners, however, were practically stripped, and wore loose breeches.

As they ran along between the horses' hoofs in a ghastly attempt to keep up with us, or have their necks broken by the sudden strain on the leathern collars, they looked a pitiful spectacle indeed. They were doing a sort of hop-skip-stumble-and-jump race, which might have been very comic, if it had not been so damnably tragic. And the only relief they had was to mouth foul oaths upon the Legion, the French officers, and the French as a nation. All in German, of course, of which "Gott in Himmel!" was a frequent interpolation.

And this was but the beginning! When we had gone sufficiently far out into the desert to please the Skipper, we were halted, and the real, serious business of punishing these two incorrigibles was begun. A number of heavy stakes were driven into the ground so that they stood only a foot or so above the sand. Then Behman and Sargas were spread-eagled in the dust, their hands and wrists being lashed securely to the stakes. For the next few hours as long as it pleased His Majesty, the Skipper, they would remain there, staring up with burning eye-balls into the fierce glare of the sun. It was our job as escort to ride round the tortured creatures in a big ring, round and round all day long at a walk-march, not too far away from them, so that they might become maddened by the sight of the incessantly circling hoofs! One horseman was allowed to break the ring at a time for a meal, and this he was compelled to take inside the ring, where the sun-blackened, parched, thirst-maddened wretches could see him.

So it went on, hour after hour, through the perishing heat of that day, while the exposed skin of those two lying there blistered and cracked. Here, indeed, was the refinement of cruelty, devised by French commanders who have studied at first hand the methods of the Arabs, Riffs and Druses, who are beyond question the world's most diabolical torturers. To add to the excruciating agonies of the two Germans, there was the cloud of dust raised by the trampling hoofs that almost smothered them at times, filling their nostrils and gritting their cracked lips. And there was the torture of the flies. Until you have been bitten by a desert fly, you cannot know, cannot imagine the appalling terror of having a thousand such creatures amusing themselves over a sunblistered chest.

The prisoners were spread-eagled at about eight o'clock in the morning, and by four of the afternoon they had passed beyond feeling. There was, therefore, no necessity to keep them any longer. Those two men, brave beyond belief, who gave no mercy nor expected any, afraid of neither man nor devil, were at long last tortured into submission. Captain Dupont was satisfied that he had broken them in. Henceforth they would, no doubt, be good soldiers, giving no trouble.

McCann and I, no children at the game, were both on the verge of giggling with hysteria as we loosened the straps and freed them. The Skipper, on the other hand, looked on with his black bearded poker face, and his air of supreme nonchalance. The two Germans rolled over in the sand as we loosened them, hiding their faces from the burning sun.

Suddenly Behman leapt to his feet with a yell, turned wildly, and started to run into the desert. Captain Dupont literally flew into his saddle and raced after him. We stood and watched. There seemed no doubt that he had gone *cafard.* The punishment had been too much for even his endurance. The Skipper's Mauser cracked once. Behman threw up his hands, then slithered into the dust. That was the end of him.

Now we were watching Sargas very closely. He had seen the shooting of his chum with wild eyes. I had a strong suspicion that he was waiting for his opportunity. The Skipper looked at him, and Sargas returned the look, incredible hate in his eyes. He was dragged to his feet, and between us we carried him back to the cells.

“That guy won't last long,” commented McCann, as we left him and returned to barracks.

“It looks to me as if he would make more trouble. I didn’t like his eyes a bit. We haven’t heard the last of him.”

“Surest thing you know, boy. That guy’s going to take it out of somebody . . . before he follows his pal. He's *cafard,* too. But he's more cunning than Behman. Somebody will need to keep his eyes skinned.”

I knew what he meant. That “somebody” was the Skipper.

The next day we learned that our suspicions had been confirmed. Sargas attempted to knife the Skipper, and was shot down in cold blood.

CHAPTER IV: TO THE RIFFS—FOR FREEDOM!

By this time we had begun to find our way around Fez. Like most north African towns, it was indescribably filthy and amazingly well supplied with drinking shops. We had sampled some of these, at least those in which the legion was admitted, and we had been thrown out of one or two in which they were not. But there was one in particular that pleased us mightily. Not because it provided any better show than the others, but because of its situation. It was an underground den kept by one Cailis, a Greek, an incredibly fat and filthy fellow who could be the soul of discretion. Other men of the Legion had found it, too, in fact, legionnaires were the only soldiers who frequented the place. For the rest, there was a motley crowd of low-class Arabs, negroes, half caste French and Portuguese, and of course the dancing girls, who seem to have originated from every quarter of Europe and the Near East.

This secret den of drink and debauchery my chum and I had named the "Pit," since that seemed as apt a description as could be devised. We stayed nearly a fortnight in Fez, which was a long time for any detachment to be at rest in any particular spot, and during this time we got into the habit of wandering, of an evening, through the tortuous alleys and back streets of the native quarter and down to the Pit.

There, in a huge room, with a bar at one end, tables round the walls, and a dancing floor in the center, we would spend many weird and wonderful hours. On a raised dais sat a wrinkled old man, whose only covering was a bright red loin cloth embroidered in fantastic designs. He strummed on weird string instruments; in fact, he seemed to spend the whole of his waking hours in this occupation, for we never went into the place but he was strumming, so that we came to associate the incessant noise with the Pit, just as people of a big city associate the noise of traffic with the city itself, and never hear it.

Those boisterous evenings usually began with pseudo dancing. A bunch of girls, some white, some olive, some brown, pretty as paint and scantily clothed, would take the center of the floor, and begin a series of posturings that could not by any stretch of imagination be termed dancing. Every one of them was a contortionist in herself, writhing her supple form into the most amazing poses, graceful yet sensuous, with certain swift and unexpected subtleties, sinuous, alluring, invitingly voluptuous, tinkling with love and laughter.

And what could battle-scarred, sex-starved legionnaires do against such feminine artistry? How did they respond to such an environment? So far as the impudent, merry-making girls were concerned, it was all a foregone conclusion. They ended the dance, inevitably, on the knees of the Legion. Then there would be much drinking with wine at ha'penny a gallon! Love-making, public and unashamed, pleasures and excitements specially distilled to appeal to soldiers of the Legion.

Brawls were frequent, knives were flashed over the possession of a particularly bewitching jezebel, when, fuddled with wine and crazed with desire, two of the boys would start a stabbing affray. This would become a sort of "extra item" of the night's entertainment. The dancing floor would be cleared. The two legionnaires would take off their heavy boots, draw their knives, and face each other, while all the girls would retire to the laps of their "friends for a night," smoking and drinking and vamping, and thoroughly enjoying the antics of the two combatants. These girls would yell their encouragement to one or other of the fighters, using half- a-dozen different languages, and promising themselves to the victor. And all the time, that fat old Greek, Cailis, would sit behind his bar and beam and beam and beam. The hoary old musician in his flaming loin cloth would strum and strum and strum.

The result would be a sanguinary mess, and no satisfaction for either side. The two legionnaires, with their cuts and slashes and bruises, would retire exhausted, to drink themselves into a stupor, for by this time most of the girls had disappeared and there were no prizes for distribution.

McCann and I had been frequenting the place for nearly a week when we noticed that something decidedly strange was afoot. About a dozen of our comrades had taken to bunching themselves together in a comer, away from everyone else. They even drew their tables together as if in some sort of conference. Their behavior was extremely secretive, to say the least; for they drank with caution, took no part in the carousals, and entirely ignored the girls! Most unsoldierly conduct!

The two principals and spokesmen of this group, a German named Hacken, and one Leondoff, a Russian, were quite well known to McCann and me. They were tough fellows and very dependable. “Say, Barrington, these guys look as if they were bent on starting a war of their own. What say? Shall we butt in? ”

I agreed it was time we knew something of what was going on; and so we went over and seated ourselves amongst them. Our reception was a sudden silence and cold unfriendly stares, but we were by no means daunted. We knew our legionnaires, and, what was more important, our legionnaires knew us. Besides, no man in the Legion carries tales. In all my years of service I knew only two men who attempted that risky proceeding, and both were found later with their own knives buried in them up to the hilt.

When the group had wanned to our presence we were told of a wonderful plan for a wholesale desertion. It seemed that information of a more or less reliable character had been passed through from Abd-el-Krim, Chief of the Riffs, that if any of the legionnaires wished to desert and would give themselves up, he would undertake to convey them safely to their own countries. McCann and I had heard of this sort of thing before. We had seen others depart and return! But the Hacken-Leondoff group would have none of our doubts and warnings. They were all set for the big effort. Rations and food had already been conveyed to a secret spot in the desert, where they could collect them and then proceed towards the Riffian mountains. Abd-el- Krim would do the rest. In fact, they were to start that very night; and though McCann and I refused to join them, they knew their plans were safe with us.

A legionnaire thinks of little else but ways and means of escape from the horrors of life in the Legion, but this seemed to me one of the craziest I had ever encountered. For these boys to move off in a bunch across the desert was the height of folly.

"They must be mad to go in a crowd like that," I said, as McCann and I returned to barracks that night.

He expressed his contempt of the whole scheme in three words:

"Some bloody hope!"

Precisely! A dozen legionnaires trudging together across the desert would most certainly be taken for a patrol, and as soon as the Riffs caught sight of them they would be shot down, murdered before they had time to explain themselves. Always supposing, of course, that they could keep clear of the Partisans while they were making the Riff country, a difficult feat for two men, let alone a dozen!

I rolled into my blankets that night with my mind full of the crazy deserters, wondering greatly just what would happen to them. For several days the Moroccan Front had been fairly quiet, and that could only mean that the Partisans would be more than usually active, since there were more attempted desertions during the quiet periods than at any other time. I felt pretty certain those foolhardy warriors were bidding for inevitable capture and punishment; but I had no idea, as I fell asleep, how soon I should find my premonition justified.

The next day Captain Dupont took out a platoon of the newly reinforced Fifth Squadron on a tax-collecting expedition. These excursions were always popular among the legionnaires. Men almost begged to be included in such parties. As the Skipper's servant and bodyguard, I was always there, and since McCann was in the same platoon, we got a great deal of this kind of diversion.

The chief thing about tax collecting was the excellent opportunities it provided for looting. Therein lay its principal attraction. Although there were occasions that offered plenty of excitement and adventure, it was the looting and the woman-baiting that mattered; and these diversions were indulged in by all ranks, from the commanding officers downwards.

Fez and its environs made a particularly good hunting ground. As the walled town itself was practically shut off from the scene of our operations the orchards of oranges, pomegranates, apples, apricots, and the olive gardens we went to work with comparative freedom. The procedure was for the whole detachment, of fifty mounted men, to ride in a body up to each farmstead and demand the tax forthwith. Obviously a good many of these peaceful, hardworking Arabians were unprepared for our advent, and not always in a position to pay the tax. The order then was to take in kind in lieu of hard cash.

In other words, the officers and men were given the chance to take what they wished, not merely in the orchards, but in the homesteads as well. Both McCann and I, who had been on these expeditions up and down the country, already possessed several hundred pounds' worth of loot, mostly antique jewelry; and as for the Skipper, that man carried his saddle bags full of it everywhere he went, and he was constantly adding to it. The Legion, on these terrible occasions, was a law unto itself.

What tax-gathering meant to the women and young girls of the unfortunate homesteads can best be left to the imagination. The example that our commander set his men in this connection was about the most gross and obscene I have ever witnessed; and, heaven knows, I saw enough 1 If I had been with him fifty years, instead of less than five, I should never have been able to look upon his exhibitions of animal passion entirely unmoved. I was no saint. I, too, was a legionnaire. But his craving for girls of school age was positively inhuman.

The Legion, of course, is a thing apart from all other human activities. The manners and customs of the men bred by the Great War were saintly in comparison, for that was a transient thing. The Legion goes on and on. It has carried on for a hundred years, and if the nations, as a whole, do not waken up to the colossal evils which it creates wherever it moves, it will go on for another hundred. It is beyond human conception. It has to be experienced to be believed.

Its logic is that only rats may be bred in sewers. The French Foreign Legion is the cesspool of humanity.

Having satisfied our various hungers, we turned our mounts towards Fez in the late afternoon of that day and made for the barracks. We were a bunch of tired savages who had adventured and feasted not wisely but too gluttonously. Every man carried a full pack in more senses than one.

We had been riding for rather more than an hour and were within sight of the walls of Fez, when we saw a band of horsemen coming slowly towards us. The Skipper called a halt, and sat surveying the strange band through his glasses. Before the party came near enough for us to see them properly McCann and I had drawn our own conclusions.

They were, as we suspected, a number of Partisans, and they brought with them seven wretched deserters. They came slowly, painfully, it seemed, for the prisoners were being dragged along on foot. Never have I seen such a ghastly sight! It composed seven of the men, our own comrades, out of the twelve who had bolted the night before. But what men! There was Hacken, the big German, his eyes gouged out and raving mad, and the rest of them mutilated in a manner too impossible to describe in cold print. Of Leondoff there was no sign. God knows what had happened to him! I looked at the gibbering wretches groveling there in the dust. Captain Dupont was questioning the Partisans. They explained that they had found the seven legionnaires in this condition at the foot of the hills beyond the plains. Some were already buried up to the shoulders and they had had to beat off the vultures in order to reach them.

I looked at the sergeant of the Partisans, saw the light of triumph in his black eyes and the cruel smile on his mouth. I suppose I was pretty well charged with the wine and other things I had thieved, and when I saw that look on the man's bestial face I suddenly went mad.

"You dirty swine!" I shouted. "You have done this!"

McCann grabbed at my arm, but I was already out of the saddle, and in another moment I had leapt with such force at the sergeant that he, unprepared, stumbled to the ground with me on top of him.

And there I started in deadly earnest to choke the life out of him. I can feel now the thick red bristles of his leathery throat in my fingers, feel the exultation that swept me in those few wild moments, see his huge eyeballs popping out of his head in a stare of utter astonishment, his thick camel lips and the huge nostril cavities. . . .

And then I felt the sudden, stinging lash of a whip across the back of my neck. It came again and again, swiftly and with unerring aim, instinctively prompting me to hunch my shoulders in an effort to cover my neck; and still I held on to that foaming, struggling devil, digging my blunt nails deeper and deeper into his leathery flesh. I've no doubt I should have finished him, there and then, if the Skipper had not started to boot me in the ribs with all the force he could muster. That was effective enough. I let go and stumbled to my feet.

"Barrington, *nom du chien!* Eengleesh bastard! *Yelia! Yelia! Yelia!"*

The Skipper and I faced each other for one long weird minute. Something like half a smile slouched across his face.

"He got what was coming to him," I said, realizing that I was out of breath. "I'll murder the swine one of these days!"

The Skipper pointed to my mount with his whip, and intimated that he would deal with me later. The prisoners were slung up with us, the Partisans dismissed, and we proceeded on a funereal march back to barracks. In the saddle with me was a Spaniard, one of the ill-fated dozen, and on the way back he told me what had happened. The Partisans had ridden them down just as they were approaching the hills. They had put up a good fight, had indeed given a good account of themselves, before they were finally taken prisoners.

Then had begun the diabolical game of torture. Two of the legionnaires had been taken out of the group, their ankles tied and attached with ropes to the saddles of two of the Partisans, who then proceeded to ride round and round a boulder-strewn patch of ground, the others being compelled to look on and listen to the screams. When released they were found to be dead.

Three others went *cafard* and were last seen racing towards the mountains. One by one the remainder were dealt with. The Spaniard whom I carried had lost an arm, or part of it, and a great deal of blood from the crudely-bandaged stump. The sight of it, in my agitated state, made me want to vomit. I heartily wished then that I had stuck to that sergeant a little longer, long enough to finish him.

For my few moments of madness I spent the night in the cells, in total darkness; but as I slept most of the time, it did not worry me much. True, it was a weird experience not to know night from day in that pitch blackness. I slept like a log for several hours I calculated, later, that it must have been ten hours' unbroken sleep. When I awoke I stared hard, trying to penetrate the blackness. It was impossible, of course. I blinked drowsily, and dozed off again. When I became fully awake I felt round the stone walls. I found the door with my hands, then sat on the floor facing it, waiting for it to be opened and the blessed light of day let in. I seemed to wait hours and hours in that pitch blackness, playing silly games with myself, opening and shutting my eyes, then keeping them closed tightly while I counted a hundred, two hundred, three hundred, opening them again with the expectation of seeing daylight. On one of these silly occasions I counted into thousands. My childish disappointment on opening my eyes was so great that I began to curse and blaspheme; but the sound of my voice was such good company that I began to sing and shout. I did one or two stump speeches, telling the whole French nation what I thought of them in good round terms. Then I began to recite. I was amazingly happy all at once, and told myself I did not care how many weeks they left me there. A scrap of Shelley came into my mind and I spoke it as loudly as I could:

[11] "Men of England!

Heirs of Glory!

Heroes of unwritten story!

Nurslings of one mighty mother! Hopes of her and one another!
Rise like lions after slumber! In unvanquishable number!

Shake your chains to earth like dew!

Which in sleep had fallen on you! Ye are many!

They are few!"

The end of every line was a startler and a pause, just like that. Well, I thought, that's good, that's splendid. Now what about something else? I did not realize then that it was a bad sign when a man talked to himself; but I had to do something or go mad.

All sorts of weird memories came into my mind. One in particular started me off again. I recalled an occasion in Baghdad years before, when a young Welsh lad and I managed to lose ourselves in the mysterious and crooked little alleys of the Arabian quarter there. We wandered about, up and down those dirty Ettle alleys and back streets for hours on end, trying to get back to the Bazaar, off New Street that wide thoroughfare which Johnny Turk had cut through the city with his cannon, from, the North Gate to the South, cut as clean and straight as a knife cuts butter. I remember that, as we tramped up and down, in and out of the maze of alleyways, trying our Arabic on the passersby, and getting sadly misdirected, I worried young Taffy into repeating several times over that incredibly silly verse about his countryman who was a thief. In some respects there was a similarity about that purposeless wandering in Baghdad and my first taste of the pitch black cells. At all events, I must have thought so then, for those stupid verses came into my mind and I shouted them out loudly and defiantly:

"Taffy was a Welshman, Taffy was a thief!

Taffy came to my house

And stole a lump of beef!"

On the word "beef" I reached a terrific yell of triumph; but, having regained my breath, I went desperately on:

"I went to Taffy's house,

Taffy was not in, Taffy came to my house and stole a silver pin!

I went to Taffy's house, Taffy was in bed.

I picked up a poker and hit him on the head!"

But I could not remember all the verses. So I did the obvious thing. I repeated all I knew over and over again. When I had thoroughly exhausted myself I lay down with my head against the door and my eyes tight shut. I had a sudden terrified feeling that it was not really dark in the cell. It was only that I had gone blind!

I began to argue the matter out with myself. I poked my fingers in my eyes till I could *see* spots and stars in the blackness, and I argued that if I could *see* those floating, shadowy spots I was not quite blind. There was still time! God in heaven! If only they would open that door I might still be saved from blindness! I began to pound on it with my fists. At that time they were pretty tough fists, but by the time I had exhausted myself my hands had burst in several places and I was reduced to a shivering sweat. After that I lay still for a long time, my eyes tight shut. It really didn't matter, even if I was blind. I should have finished with the rotten Legion, anyway.

I developed a marvelous scheme then. I could *see* myself being discharged, being escorted to the train for Damascus, then taken aboard the ship for Marseilles, escorted by train to Paris, shipped to England, and finally dumped in London. I was not clear how I should make my way to Trafalgar Square, but I could *see* myself there, on a soap box, telling a very harrowing tale about life in the Legion, and incidentally my blindness. I was just in the act of taking a thumping big collection, when I heard the rattle of keys and the door being opened. I scrambled to my feet . . . and met a sudden, blinding flash of daylight!

Of course, I was blind, blinded by the light of day, but not at all in the sense I had been raving to myself about. I staggered out into the corridor, my hands groping along the walls. Two of the jailers grabbed me and hustled me along, out of the prison, across the courtyard, where, at the gates, I was handed over to an escort of the Legion, and then marched off to face Captain Dupont. Gradually I gained the full use of my eyes, and by the time we had reached his quarters I was very nearly normal again.

The Skipper met me with a mouth full of curses, but refrained from using his heavy boots. I started to clean his kit, tidy his room, and generally assume my multifarious duties. It was, I saw, about four o'clock in the afternoon, so I must have been in that infernal cell for about twenty hours. As I say, it did not worry me a great deal, except for the last two or three hours. I had yet to learn what it was like to be in that pitch blackness for seven days.

As I went about my job I happened to catch the Skipper's eye on me. I took the opportunity to tell him that I could do with something to eat and drink, and I was booted for doing so. That did not worry me either. I could see he was merely preserving his attitude of stolid brutality. One got accustomed to those kicks in the belly for the Skipper could lift his foot quicker than most men could lift a hand and it became a habit to harden the diaphragm in anticipation. I don't know that I could stand such kicking now. In the Legion it is different. You just adapt yourself and so become carcass-hardened. It is either that or going under with a terrific amount of suffering. A sort of survival of the slickest. I repeated the Spaniard's story of mutilation to him, and with amazing effrontery suggested that the Partisans should be made to pay for their dastardly work. After all, they had destroyed twelve good legionnaires.

"Twelve dogs!" snarled the Skipper. "We should have lost them, in any case. What the Spanish swine says may be true, maybe not. But whatever happened, they asked for it. . . . A lesson to you, and to others."

"Mon Capitain," I persisted, "will you not take action, and punish these inhuman beasts of men for their horrible crime? Our men will speak in evidence."

He laughed. *"Non! Non! Non!* It cannot be. These are cases where the word of a Partisan is taken before that of such dogs of the Legion. Partisans are good soldiers. They are useful to France."

And that was the last time I ever attempted to rouse authority in the cause of justice where the Partisans were concerned. The Skipper dismissed me with his foot, and told me to get to my own quarters, where I should find some food waiting me, ending with a warning that I must keep my nose out of the affairs of the Partisans in future, if I wished to continue as his servant. I realized, of course, that I had been let off lightly with only twenty hours' cells, but that the

Skipper had to do something to save his face before the Fifth Squadron, who both feared and loved him.

When I got back I found McCann awaiting me with enough food to feed half-a-dozen men. His India-rubber face creased into a delightful grin when he saw me.

"Kim on! You son of a gun! Dig into this, Barrington, and then we'll go out and paint this goldarned city black and blue! "

A great pal, McCann! In his company the brutalities of life in the Legion did not seem half so bad. I wonder if he or I could have endured the life as long as we did if it had not been for our wonderful comradeship? I very much doubt it. Six thousand miles of ocean divide us now, but I shall never forget, nor will he.

"Heard the news, boy?"

"What news?"

"The Fifth are to move up to the fort of Ammasine."

"Good! So it was *that!* I thought something had happened to put the Skipper in a good humor."

"Yeah? Waal, we shall find a bit of scrapping and get back to normal, I reckon."

That night McCann and I swore a vendetta against all Partisans. We were determined that, when the opportunity came our way, we should give them as good, or as bad, as they had given us.

I did not dream then that, many moons later, I should be put in chains for shooting twenty-four of them!

CHAPTER V: THE LOST GARRISON

On the march up to Ammasine, McCann, to his intense disgust and my delight, was placed in charge of the mule transport. It proved to be a great source of amusement to the Squadron generally, for they could never understand why, since Mac and I were such inseparable pals, we were perpetually ragging each other, not to say quarrelling. The newcomers, especially, found it difficult to appreciate a comradeship that appeared to consist largely of a fire of blaspheming backchat and repartee. Our attitude convinced them beyond any possibility of doubt that these English-speaking people were mad, quite mad.

Nevertheless, McCann's antics in trying to keep the stubborn and stupid mules in line were really funny. There was little he did not know about horses, but he hated mules with a hatred that was beyond all reason. And he showed it in his handling of them. Now, a mule is a very cunning animal. He has far more intelligence than he is generally credited with; and he has a way of his own in responding to dirty treatment. As I watched McCann, occasionally throwing out hints and generally helping to make matters worse, I felt certain he was in for trouble.

It came. I was riding leisurely along when I noticed a confusion among the kicking mules, and heard a bawling American voice. The mule transport had got itself tied up in a frightful mess. I rode over and found McCann valiantly endeavoring to get two of the animals to rise. One of them was carrying a barrel of the troops' wine, and in going down it had struck a boulder. In an attempt to stop the flow of precious liquid Mac put his mouth to the gap, but it was too much even for him. He tried to hold it back with his hands and was pretty well drenched in the process. He looked as if he could have wept at the sight of all that wicked waste. And imagine the anger and pain in his eyes when I asked him what was the matter.

"Matter! You damn' limejuicer! What in hell are you sitting up there and grinning for? Oh, boy!'' he bleated, '"can't you see what we're losing! Give us a hand, for the love of Mike! " There seemed to be no thought in his head for the fact that he was holding up the whole column with a couple of mules. It was the wine, the spilled wine, that made his heart bleed. When Captain Dupont came up and saw what had happened, the two mules down, the wine leaking, and poor old Mac vainly trying to stop it, he stared incredulously. It looked for a moment as if he were about to use his whip on McCann, but, seeing the funny side of the picture, he started to roar, and so did the rest of us.

Some of the boys got down and helped to put matters right, and we continued our march with the loss of nearly half a barrel of wine. McCann was never again placed in charge of the mules. For the remainder of that trip he simply groused and grumbled because it had taken "a bunch of those lousy guys" to help him stop a leakage of wine. For this terrible offence he was docked ten days' pay, which amounted to five- pence, since a legionnaire's pay is only one souper day during his first eighteen months of service.

"We don't draw much from the lousy frogs as it is," he commented, "without trying to stop a guy's pay. I guess, Barrington, you should come fifty-fifty with me on this, you big hobo."

"Not on your sweet life, boy!"

McCann grinned. "Waal, I reckon I got rid of the mules, anyhow!"

We were still within some eight miles of Ammasine when the escorting column withdrew and returned to Fez, leaving the Fifth Squadron to proceed alone. A little while later we came within sight of the fort. We saw the French Tricolor flying over the outpost. We drew closer and rode to attention, anticipating the appearance of the Commandant whom we had come to relieve. We drew nearer and nearer, but still no one came out from the fort to meet us. We were now within three to four hundred yards of the road leading to the fort, but there was no sign of life anywhere.

This was decidedly strange, to say the least. Captain Dupont called a halt, and gathered his officers round him to discuss this most unusual situation. Then he called a corporal, McCann and me, and ordered us to proceed up to the fort, taking every precaution. Frankly, we did not like the idea at all. There was something altogether mysterious and sinister about this silent fort. What had happened to the Legion infantry who should be in garrison there? We rode cautiously up to the main gates. We were not afraid of standing up to anything we could see. But there was too much of the unknown about this ghostly silence. If the place had been taken, why was the tricolor fluttering in the breeze up there? We reached the gates, hammered on them with our rifles, and yelled:

"Ouvrez la porte I Ouvrez la porte!"

But there was no answer. Not a sound.

"Hell!" cried McCann, "these guys must be all liquored up!"

The corporal suggested we should climb on to the wall. We drove close up to the masonry, stood up in the saddle, and climbed to the top. I don't know what we expected to see when we looked over, but there was not a soul to be seen anywhere, everything was deathly quiet. What was the secret of this mysterious silence brooding over the fort? We cleared the walls, and landed inside. Still not a sound, not a movement. We opened the big iron studded doors, and then signaled to the Squadron waiting below.

Captain Dupont, fuming with rage at the unaccountable delay, came riding up at the gallop. We explained the situation to him. He at once detailed search parties to go through every cranny and comer of the buildings of the fort, while he remained in the central parade ground, pacing up and down, and pondering on what could have happened to the force we were relieving.

Everything was in perfect order. In the barrack rooms beds were carefully laid out. The stores and the magazine room were clean and tidy, and still contained a quantity of ammunition and rations.

The only sign of disturbance about the whole place was the fact that the telephone wires had been cut. That was significant enough.

It was the strangest thing I have ever experienced, in its way, the whole fort in apple-pie order and not a soul left in it! There was not the faintest sign of dead or wounded, not a scrap of leather, a gun, a boot, or even a button, to mark the presence of those who had been there. The place had a distinctly unpleasant atmosphere, and the men were conscious of it. They were restive. There was talk in the ranks of the fort being bewitched or haunted. As first one party returned to report to the Skipper and then another and another, the tale was the same nothing. By now there was a distinct rumble among the ranks. Men were asking what the Captain intended to do. And he answered them pretty quickly.

“Do! You *sales cochons!* You ask me 1”

He snapped out the order to fall in, and threatened to shoot any man who dared disobey. The ranks stiffened to attention. Once again they were just men of the French Foreign Legion. Double sentries were posted on the walls, and on the various rooms, with a special guard on the magazine room. Then the Skipper called the officers for a conference in his headquarters. After discussions that went on for a couple of hours I was called in to prepare the Skipper for the night. Looking at me from beneath his bushy brows, he said:

"What do you make of this, Barrington?"

“I think, *mon Capitain,* I might be able to solve this mystery if I were given a free hand, and time . . . twenty-four hours.

I have an idea at the back of my head. If you would permit McCann and me. .

"Very well, bring in McCann."

I found my chum at his usual pastime filling up with liquor.

"What does the damned hobo want now?" he groused.

"Can’t he leave a guy alone for five minutes?"

"This is serious, Mac."

"Yeah? Serious? Those guys left some good wine behind, wherever they went. Fancy them doing that!"

Several other officers had gathered in the Skipper’s quarters when McCann and I went back. Some of them looked uneasy, others a trifle bored, but they all looked up as we marched in.

"Now, what is this wonderful idea of yours, Barrington?" growled the Skipper.

I suggested that whatever had happened to our comrades of the Legion infantry, the Riffs were responsible for it. It was some trick of theirs. They had in some way spirited the men away . . .’’

"You're crazy!" interrupted McCann. "How could those guys be spirited away? The only spirits around here . . ."

"Shut up, McCann!" roared the Captain. "Exactly what do you mean, Barrington? ''

"Sir, I think there must be some secret way out of this fort, a way known only to the Riffs. I should like a free hand to investigate." Turning to the officers, Captain Dupont asked their opinion.

"Les Anglais sont foux!"

And there was a roar of laughter at my expense. But I still persisted in my theory. I really believed I could solve this mystery. At last Captain Dupont consented, and McCann and I left the room with twenty-four hours at our disposal, during which time we must find how the Garrison was spirited away.

Outside, McCann looked at me in a mournful sort of fashion.

"Poor old Barrington! I never thought you would get it so bad, kid. D'ja get any pain with it? What are the symptoms? You poor bloody fish! To think you could solve a problem like this. Boy, I always thought you were crazy. Those officers were right. *Les Anglais sont foux!"*

"Put a sock in it, Mac. This is serious. In the next twenty-four hours we're going to examine every inch of this benighted fort."

"Yeah? O.K., Keylock Holmes! Lead on! Let's *go to it."*

And how we searched! Hour after hour we wandered about, examining every nook and crevice, the mad American and the still madder Englishman! Every man in the Squadron looked on, tickled to death at our antics. During those painful hours we were the butt of the whole crowd. They pointed at us and tapped their heads significantly. I soon realized that if I failed in this project I should henceforth be the prize idiot of the Squadron. I felt rather sorry for having dragged McCann into it. Not that he appeared to mind. He took advantage of the situation to cadge bidons of wine at every opportunity, and since our progress was a source of amusement to the rest of the boys, he managed to collect a gallon or two during the search.

We came at last to the magazine room, a stout stone structure. The sergeant in charge of the guard walked round with us, freely expressing his opinion of us and our quest. He said he had never seen anyone go *cafard* in quite the way we had done . . . and so on, *ad libitum.*

! Round the walls of this room were stacked cases of ammunition. I decided that to have them all removed would serve the double purpose of helping me in the search and getting my own back on the sergeant. He disliked the idea intensely. It meant putting all his men on the job, and he did his utmost to persuade me to search elsewhere. But as I carried orders from the Skipper, he dared not refuse. And so began the task of dragging the cases into the center of the room. The men had got half-way through this task when I noticed they were having the greatest difficulty in moving two particular cases. In fact, they could not move them at all. McCann and I stood by and watched, thinking this a ruse, an attempt to give up the work as hopeless. They left the two cases there eventually, and went on to remove the others.

When the whole of the walls had been cleared I pointed to the two cases that had been left, but the guard shook their heads, pointing out that those two must be stuck to the floor in some way, since they could not move them. It was then that McCann and I looked at each other with slightly raised brows.

We tugged and pulled at the cases ourselves, but they could not be moved. At last we decided to open them and take out the contents. When this had been done we found that they had been secured to the floor of the magazine room by means of hand-bolts screwed home from *beneath* the floor. Strangely excited at this discovery, I had the cases taken to pieces, and as we freed the lower boards we saw that they had been used to cover an aperture in the floor. Beyond this aperture were steps leading down into the bowels of the earth.

I immediately sent one of the guard to report our find to Captain Dupont. Another was dispatched for hurricane lamps.

In a few minutes the Skipper had arrived, accompanied by two other officers and a number of the boys. There was no need to explain anything. That cavity spoke for itself. It was not an old dungeon. It was new. The place smelt of newly-turned earth, which was very significant indeed.

The Skipper took a lamp, drew his Mauser, and started down the steps. We followed. At the foot we came into a large room with earthen walls reinforced with timber, and around them were cases of powder with a fuse laid all ready to blow the fort sky high!

Leading out of this chamber was a long passage which seemed to descend deeper and deeper as we passed through it. Suddenly it opened out into a wide chamber several yards in extent, and the Skipper, who was leading, raised his lamp and called the order to halt. Before him was a huge expanse of water. We approached cautiously, and to our horror, found the water filled with dead bodies.

"S acre bon Dieu!" cried the Skipper, "what is this?"

We crowded round the edge of the water, then found we could splash through it, for it was but a shallow Pool of Death we had stumbled upon. The stench was appalling. Rats leapt. We had to step with extreme care in getting across, since those mutilated bodies beneath our feet were once our comrades of the Legion.

Messages were sent back along the tunnel for fatigue parties to get busy. I did not envy them their job, and I was immensely relieved to reach the other side and continue with Captain Dupont along the subterranean tunnels. Always it seemed that we went down an incline, down and down as we progressed, and presently we saw ahead of us a tiny spot of light. It grew bigger as we approached it, and presently we came to another chamber. This was well stacked with cases of ammunition. Here the air was much sweeter and cleaner. We rested while Captain Dupont went forward cautiously along the few remaining yards of tunnel. We watched him approach the end of the tunnel, saw him drop to his knees and crawl the last few feet towards the outlet and peer out.

In another moment he was hastening back to us. We jumped to our feet, guns ready.

"Mon Dieu!" he cried hoarsely, "but there is a great army of Riffs out there, beyond the ridge, in such ambush that they cannot be seen from the top of the fort! Come!"

We followed him and looked out. There, in the valley, was an encampment of Riffs about three thousand strong. They had thought their undermining of the fort safe from detection by securing the entrance to it in the magazine room, and so, fortunately for us, they had not troubled to leave a guard at this end of it. But Captain Dupont took this precaution, and the remainder trooped back to the magazine chamber. Presently a party of men came down to us with rations and a plentiful supply of wine.

By this time everyone had changed their attitude towards the mad American and the still madder Englishman. We were, indeed, privileged men, eating and drinking while others slaved to carry out the Skipper's orders, and we enjoyed ourselves thoroughly. Naturally I felt proud of my achievement. I do not know even now what prompted me to pursue that seemingly foolhardy quest. I had argued that since the Garrison had been spirited away in some manner by the Riffs, without leaving a trace of any kind behind, since the Riffs had not then taken possession of the empty fort, it was clear that whatever means they had used would be employed again, after the relieving force had taken over. In short, they had some secret means of getting in and out of the fort. It was possible that they would use the same method to capture us as they had used before. Having done this once, was there any reason why they should not do it again, cleanly and as mysteriously, again and again? In this way, they might E account for a great many companies and squadrons of the ill-fated Legion before being discovered.

I do not know what the orders were that caused us to be sent to relieve the fort at Ammasine. Until I came up against this mysterious problem I had thought we were relieving a garrison which was due for a rest. Then I began to see that it might well be otherwise. We had been sent to relieve a fort from which there had been no news for days. Supposing, therefore, we were spirited away, too! Very well, then, another squadron of the Legion would be sent along to see why we had not communicated with headquarters at Fez since our arrival . . . and another, and another!

I could see this trick being played quite a number of times, before a vast column of the French Army would be sent along to investigate the whole mysterious business! There was something about it that smacked of the Riff mentality. Abd-el Krim was not merely a powerful chieftain, he was immensely wealthy. Nothing was too big for him. As I have shown, he could command the expert assistance of all manner of specialists in warfare. His agents at that time must have been flung right across Europe, for he had bought big guns in Britain, and had bought the help of Europeans in organizing his vast hordes of fighting men.

I do not know what happened in other lonely outposts at this particular time. I can only speak of my experience at Ammasine. But I should not be in the least surprised to know that other French outposts had been undermined in the same way by the Riffs, working in large batches, and that other garrisons were similarly spirited away. Imagine the possibilities along the whole chain of these forts of the Moroccan Frontier! It should be remembered that Abd-el-Krim's campaign on this occasion was laid along the whole Frontier right down to Spanish Morocco. Think of the splendor of such a scheme of undermining forts along the whole front, and carried out simultaneously by well-equipped engineers.

Looking back on the strange episode of the Lost Garrison, that is how I see it now. Whether such attempts were made upon other forts, it is impossible for me to say. Such facts are hidden away in the French records, and we may be very sure that the world will never know! All I do know is that Abd-el-Krim spent enormous sums in a gigantic effort to rid Morocco, once for all, of the infidels. We had evidences of this everywhere we went, in the big guns and ammunition we captured, and in the expert engineering fortifications that we came upon from time to time.

Therefore, it did not seem to my mind a foolish idea that there should be some secret way in and out of the fort at Ammasine. Discussing this experience on many an occasion afterwards with McCann, he did not agree with my reasoning. He argued that, even if the Garrison were spirited away so as to leave no trace, it could have been done from without the fort itself. They could have been attacked in over whelming numbers and the dead and wounded carried away. After which the whole fort could be cleaned up to give it that air of mystery which brooded over it when we arrived.

The possibility occurred to me, of course. But as soon as I began to examine it, I realized how faulty it was. In the first place, the garrison would not be beaten so quickly from an attack outside the fort that they had not time to communicate with Fez. All our experience showed that it would take several days to capture such a fort, however big the invading army, and during that time the news would have been passed on to the headquarters at Fez and a relief column would have been sent out. A small fort manned by a handful of men can be held for a long time, providing the defenders are well supplied with water, rations and ammunition.

No. There must be a secret way into the fort. I set myself the job of finding it. And I succeeded. That was the great thing. To picture the rest was not a very difficult matter. Riffs, like their countrymen, the Arabs, can move about as silently as cats. There was no doubt that a number of them had made their way to the magazine room of the fort, had overpowered the guard, and had then changed into the dead legionnaires' uniforms. After that, there would be a quick dispatch of the sentries on the fort walls a swift, silent stabbing of unwary backs. Then the crowding up of more and more Riffs by way of the magazine room, and the remainder of the Garrison would be taken before they were properly awake. That could not have proved so gigantic a task for cunning Riffs, even if they had had a full company of legionnaires with whom to deal.

From that date onwards, throughout the whole of my career in the Legion, Captain Dupont was continually pestering me to take a non-commissioned rank, and I as persistently refused. He bullied me, ragged me, brutally ill-treated me, kept up a perpetual barrage of invective and abuse, all with the sole object of forcing me to realize how much better a time I should have in the Legion if I were a corporal. He could never understand why I continued stupidly to refuse promotion, and I never told him the truth. Had I done so he would have found another servant and treated me hence forth as an imbecile.

The reason why I refused promotion is not far to seek. All noncommissioned officers of the Legion, besides holding rank equivalent to commissioned officers in the British Army, have to be experts in the art of flogging. I am not in the least sentimental when I say I am an Englishman, and could not possibly have learned to flog men as they were flogged in the Legion. It wasn't because I had not the stomach to do it. I had in those days. But I could not undertake such work because of the inhuman and brutal way it was carried out. To talk of flogging is to bring to the average mind a picture of a man stripped to the waist, his hands tied to a whipping post, and another man wielding a salted whip until the exposed skin is lacerated with wounds.

In the Legion a flogging meant much more than that. It meant a peculiar handling of the whip, so that it curved through the air and tore, not merely at the back of a victim, but at more susceptible parts. If there is such a thing as cultured flagellation, then the N.C.O.'s of the French Foreign Legion and the officers too are certainly cultured in the bestial system. I was as brutal as many another legionnaire, I have no doubt, since it is a life that brings the very worst out of a man; but I am English, and I could not learn to flog a man so that, with every stroke, there was the possibility of ruining his physical structure for life!

CHAPTER VI:
STRATEGY BECOMES A TRAP

Captain Dupont, after a conference with his officers, decided to place a strong guard with machine guns at the mouth of the tunnel. The whole length of it, right up through the magazine room to the headquarters of the fort, was wired and a field telephone fixed up at each end. The plan of campaign was to work the tunnel mouth simultaneously with the operations in the fort above. McCann and I were posted with the party at the tunnel exit, and I was in charge of the telephone.

Realizing that the Riffs were only waiting for an opportunity to attack by way of the underground passage, and so surprise the new force as they had the old, the Skipper speedily got to work. Every man was posted for duty, fully armed, and the new garrison of 250 men waited in high glee for the Riffs to make their surprise attack. It was a fatal blunder, as will be seen later. But, at the time, we were chuckling at the thought of how we were to surprise the enemy, who had planned so well and worked so hard to surprise us! It was to be a turning of the tables with a vengeance.

The Riffs were apparently in no hurry. No doubt they imagined their scheme secure, and that being so, it would be a good plan to allow the new garrison to settle in, give them the impression that the Ammasine zone was quiet and peaceful, and thus make doubly certain of their secret maneuver!

Our wait was a lengthy one. All night long Captain Dupont, who was with the men in the fort above, was continually sending messages of inquiry. But there was nothing to report. The Riffs in the valley beyond the mountain ridge remained quiet. Dawn came, and still no movement on the part of the enemy. We knew they would not advance during daylight. The long wait rather took the edge off our anticipations of the surprise attack we were to make. The band of watchers at the tunnel mouth were relieved, but McCann and I continued to take duty in turns at the telephone. Captain Dupont came down to investigate for himself. He was by this time fuming with impatience. He insisted upon McCann and myself remaining at the telephone post for another night, and remarked, in his own inimitably brutal fashion, that, since we had found the subterranean passage, we might well be left in charge of it.

However, it was felt that this coming night would find a definite move on the part of the enemy, and we began the silent night guard with renewed anticipations. They were justified. Towards midnight we saw the slow, stealthy moving shapes advancing towards us. As far as we could see, only a small band had moved out from the encampment. On they came, slowly, cautiously, like some strange animals of the night, their bellies to the ground.

"Come on! You cursed tiger-cats!'[1] growled McCann, in a hoarse whisper.

I was at the telephone, reporting their every movement. As the band of crawling men drew nearer, nearer, we noticed that another band had started out from the camp in exactly the same manner.

With my eyes on the moving ghostly shapes, I kept up a running commentary into the 'phone mouthpiece:

"First group about eight hundred yards away. Still moving, but slowly. Advancing in good order. Second group now well out of encampment. Moving at same rate as first group. Another party left the camp. Advancing slowly and in the same order. First party now seven hundred yards away. Three groups moving in column order. Second and third parties closing in a little. Another band of Riffs leaving encampment. Advancing in same order. Four parties now advancing. First party within six hundred yards . . ." And all the time I was receiving the message:

"Garde! Garde! Garde! Garde!'[1]

We were fuming with impatience by this time, especially McCann and I, for we had been on the stretch more than twenty four hours. The Skipper, of course, wanted as many Riffs out as possible, wanted them to get right on top of us before we opened fire on them. But as we watched the enemy draw nearer, almost within hailing distance, it seemed to me he was cutting it rather fine, though we dared not fire without the order.

By this time, more and more parties had left the encampment, until the plains below us seemed to be a mass of moving shapes, crawling along like animals, low on the ground and ready to spring at any moment. It was then that their movements needed a lot of watching. Finally, I was reporting the perambulations of a dozen different groups. They would make such easy targets! We could not miss them, and we itched to make a start, to spill death into them with the machine guns.

The first group was within a few feet of the tunnel mouth when the order came through. After all, it was beautifully timed. As we opened fire we could hear a faint rat-tat- tat-tat from above, which told us that Captain Dupont had also opened fire. It was a surprise attack with a vengeance. The Riffs were completely taken aback, and before they had recovered from the shock, we had demolished the first party and were pouring forth our deathly message into the second and third and fourth groups.

Shrieks and yells rent the air as the Riffs found their ingenious scheme had failed. Already their dead were lying around in heaps. But after the first shock, they came on, up the slight incline of the plains, taking advantage now of scrub and boulders, any bit of cover they could scramble into.

There was a sudden burst of life in the encampment, guns were being drawn up, the massed hordes were turning out at the double. The strategy was over. It was to be open fighting. The enemy was advancing in vast numbers. Soon, they were coming up faster than our machine guns could deal with them. They were throwing men away in all directions, for the Riffs never spare themselves once they get into the heat of battle. The machine guns were in position behind their walls of dead. Bullets came spattering against the earthworks of the tunnel.

“Gee, boy! What is a hundred men to them! It’s just sheer murder! Shooting ’em is like taking lollipops from a kid! Easy! D’ja ever see anything like it, Barrington? It’s a real war. ...”

And it went on like that for upwards of two hours. Grimly we mowed them down. Every one that fell meant we were avenging the ghastly deaths of our comrades. We saw the situation was changing, and we fought like men possessed of devils. They were bringing up their big guns and we knew what that meant.

I telephoned the position. We were ordered to hold on. A shell burst somewhere overhead, then another and another. Their machine guns were spattering a hail of bullets about the tunnel mouth, but so far they had done nothing more than pit the walls. Our little party was intact. Then more shells came shrieking through the air. Suddenly there was a deafening report immediately above us. A shower of earth and stone came toppling all about us. It threw our little party into confusion. Men, half buried, fought, struggled and cursed in their efforts to extricate themselves. Our machine guns were smothered in debris and out of action. Luckily for me, I was still at the 'phone some three yards or so beyond the tunnel mouth. I reported what had happened. The entrance to the tunnel was almost blocked up. The order came to retreat into the fort and blow up the tunnel as we went. Our men were all right, except for a few scratches and bruises. We had been in good cover the whole time.

We made preparations for blowing up the tunnel, and were joined by a fatigue party just as we started to drag the kegs of powder into position. We retrieved the guns and carried them back, laying the fuse as we retreated, back and back. When we were well out of range, the fuse was lit and we let her go. Then we took to our heels along the tunnel, splashing through the Pool of Death, kicking up the squealing rats, laughing and shouting as if we were indulging in a Fifth of November party.

Suddenly there was a deafening explosion, followed by a rumbling roar along the tunnel. Some of the men were hit by falling fragments, for the shock along the tunnel was greater than we had bargained for. But there was no doubt that we had very effectively blocked the entrance. It would take the Riffs many hours to open it up again, provided they were foolish enough to try. "Say, Barrington!" roared McCann, "I guess some of those swine managed to get into the tunnel before the fireworks."

"I expect they did!"

But I was thinking of the false strategy in manning the tunnel at all. It had very nearly become a death trap. It was beginning to look like a terrible blunder to me, and I fancied that by now the Skipper had come to the same conclusion. We should have blown it up as soon as we had discovered it, instead of wasting precious hours posting men and guns there. If that had been done, we should have been able to concentrate our efforts on the job above, with the rest of the crowd.

After all, we had seen something of the size of the enemy's camp, and should have guessed they would be armed with big guns. We had had a taste of those guns on more than one occasion. What chance had an earth tunnel against a well-timed and well-directed shell? But, as McCann said, it was easy to be wise after the event 1.

When we reached the fort and reported to Captain Dupont, we found things very lively indeed. Shells were whistling through the air. We had lost a big number of men. Our wounded lay unattended, groaning and cursing. There was triumph in the yelling of the oncoming Riffs. Suddenly the big guns ceased fire. That could only mean one thing. They would make a determined rush for the fort.

"Mon Diezi I There must be ten thousand devils storming the walls!"

There undoubtedly were! It was touch and go for the fort and the garrison during that terrible half-hour or so. There was a vast crowd of them almost on the gates, and as our machine guns vomited hot fire into the black mass it was amazing to see how the horrible gaps filled as quickly as they were made, how the dead stumbled, slithered, rolled and toppled down the slopes. Again and again we drove them back when they were within a few feet of the walls. Abd-el-Krim must have lost several hundred men in that insane attack.

Our men were thinning terribly. We had lost four officers and several N.C.O.'s. I saw McCann tear open his tunic. His thick woolen shirt was clotted with blood. He stuffed a field pad under it, buttoned up again, and went on swinging his gun, his mouth foaming with curses the whole time.

"Stopped one, Mac?" I shouted it, not daring to look towards him again.

"Clip in the shoulder. . . . Hell of a way from being finished. . . . Sons of bitches. . . . Kim on! Kim on! You big, black-faced . . . !"

As if they needed any invitation! They came on, all right, yelling and screaming like fanatics the whole time. Every battle a religious one to the intrepid Riffs. Allah came to every war of theirs. And they said so, in leather-lunged yells and shrieks. They are maniacal fighters, but if they were to stop screaming during an engagement, they would soon be finished! But they did not. They kept Allah by them. Fight or fall, it did not matter to them, Allah was there in either case. Hundreds passed out with his name on their lips, and thoroughly enjoyed themselves in the process. There was not much of the Western-Front science about the Riffs. In action, they are just madmen, and ten thousand madmen must be faced to be believed.

Suddenly the firing eased up. The enemy drew off. Hundreds of them careered back over the plains like a school let loose. We ceased fire, wondering what this ruse might be. After all, they had been within an ace of taking the fort. Why this sudden and mysterious retreat?

I found McCann propped up against the wall alongside his gun. He seemed to be enjoying himself, for a number of empty bidons collected from his dead comrades lay at his feet, and he was roaring drunk.

"Hey, Barrington!" he cried thickly, "this is a great war. The more I drink, the more I love to fight these guys?'

There was a mad light in his eyes. His face was caked with sweat and dust. I wondered for a moment whether he were going *cafard.*

“Come, Mac,” I said, “pull yourself together.”

Evidently it was only lust of battle in his eyes. He grinned as he held a bidon towards me.

“Kim on! Have a drink, you great big limejuicer! I’m all right.
I’m sure enjoying this lil’ fight!”

I shook my head. “I’m going to get a bucket of water.”

His eyes grew round in his head as he stared his amazement.

“Water! My Gawd! What for?”

“To dip my head in, of course.”

“Barrington, you big son of a gun, you’re sure going *cafard!”*

It was only a little after sunrise when we had left the underground passages, and since then we had stood up to the Riffs for the better part of the day. When the lull came I was beginning to feel as if the intense heat of the sun were cracking my skull, and it was a relief to watch it disappear beyond the rim of the distant hills.

All firing had now ceased, except the occasional phut of a Riff sniper, who was watching for movement about the fort walls. During that night there was no fighting, and no sleep either. What few of us were left found ourselves posted on the walls, and we took food where we stood. Down below us the enemy collected their dead. We did not disturb them, being content to wait for the dawn. And with the coming of dawn a most amazing sight met our gaze. Never have I seen such a force massed together. Without a doubt, Abd-el-Krim had started his long-threatened big push. It looked as if the whole Riff tribes had come out into the open with the fixed intention of finishing us off, once and for all!

A party of horsemen came galloping into view and made towards the fort, holding a white flag aloft. Evidently the Riffs wanted a palaver. When they were within speaking range, Captain Dupont asked what they wanted.

They gave us to understand that if we would surrender we could march out of the fort with full honors of war, and they pointed significantly towards the vast hordes waiting for the word to advance.

Our Commander replied without a second's hesitation:

"You and your rebel Chieftain can go to hell! When you have taken that flag down you can take the fort. Yelia!"

There was a roar of cheering from the small band of legionnaires who were still on their feet.

"Vive le Legion! Vive le Legion! Vive le Legion!"

The same, the self-same legionnaires who habitually cursed the Legion, cursed the day they joined, cursed France, and all that France stood for! Truly, a weird body of men, but lovely fighters!

Once more we prepared to sacrifice ourselves in a hopeless engagement. The truce party had barely got back to their column, when the big advance started. The enemy came across the plains in such massed numbers that they seemed like some gigantic black wave sweeping across the land. We met them in grim earnest. The roar of Allah was upon us. On they came, and though we were mowing them down like blades of com, we seemed to have little effect upon their massed ranks. The sea of screaming madmen rolled up, ever up, towards our last hopeless stand. "Gosh! Barrington, did you ever see such a goldarned mob as this? Where in hell did the old Krim *Diable* pick 'em all up, boy?"

Suddenly there came a bursting of shells but not on the fort! Those shells, and the bullets of machine guns, were dropping among the Riffs from their rear, scattering them in all directions. The massed ranks were being broken up in confusion. They fell back from the very fort walls as our relief column broke through and came dashing up towards us. All the yelling now was inside the fort. Four squadrons of Spahis Tunisian galloped up and crowded into the fort under command of Colonel La Reine.

"You must get your men together," he ordered Captain Dupont. "The whole Riffian forces are massing against you here. We shall have to make a running fight for it. Hurry! There is no time to hold council!"

CHAPTER VII: TRAGIC RETREAT—BACK TO FEZ

Of the 250 men who comprised the Fifth Squadron when we took over the empty fort at Ammasine, only ninety were standing when we left it. We collected our baggage and saddled our horses in an astonishingly short space of time. Our wounded were placed on stretchers slung between mules, and sent on ahead to join the remainder of the relief column awaiting us down on the plains.

In that tragic retreat we had, besides the four squadrons of Spahis Tunisian, numbering a thousand men, a battalion of Senegalese, two companies of the Legion infantry, and two battalions of French 75's. It was the most amazing retreat I have ever seen, heard of, or read about. It will not go down in history, because anything the Legion does is never officially recorded. Doubtless there is something in the French records of how the Algerian cavalry and the Senegalese and the French 75's fought in that memorable retreat from Ammasine to Fez but there will not be anything there of the charges, the hand-to-hand skirmishes, and the bloody battles fought by the two companies of Legion infantry and the wretched band of ninety men of the Legion cavalry, who covered the rear of the whole of that retreat, covered it until they were left entirely on their own by the fleeing, so-called relief column!

We had scarcely got down the hillside preparatory to joining the main force, when the Riffs opened fire. By this time the enemy had ambushed behind boulders, in nullahs, and under such breastwork as they could hurriedly throw up; while the big guns were sending out shells from their encampment beyond the mountain ridge.

We fell in with the column, ninety battle-worn horsemen of the Legion, and were joined by the Legion infantry, to form the rearguard a regular hell-corner when the enemy is closing on a column! The sun was getting up its terrific heat, and the day looked thoroughly promising. Already the Riffs were harassing our flanks. We dared not turn back for our wounded now. God knows what happened to them!

As the day advanced, so did the enemy. They amused themselves by making short rushes at us, yapping at our heels like hungry beasts thirsting for blood. Our infantry poured withering fire into them, and as they drew back our little band charged at them, cutting off the stragglers. Then, about turn, and away we would race back to the column. In one of these insane charges I was engaged in a hand-to-hand scuffle with a huge black-bearded Riff, and before I could finish him I felt a stinging pain in the calf of my right leg. A bullet had shot right through the flesh.

We retreated in that manner for the whole of the day. Night fell. We of the cavalry went into the center of the column on account of the horses, but the Legion infantry remained at the rear, digging themselves in as best they could and keeping up their fire, backed by the 75's, the whole night long. There were big losses that night. Men were falling everywhere. Some were shot while in exhausted sleep. We still had a score miles before we could make Fez.

Dawn came, and with it a renewed attack. We saddled our mounts, and joined the Spahis Tunisian in a mad charge to drive the Riffs back. My leg was stiff as a board, and excruciatingly painful. But it was a case of stick it or fall into the hands of the Riffs.

Suddenly my horse reeled under me and fell gasping to the ground. I managed to throw myself clear, dropping safely on my uninjured leg. In the general melee that followed I grabbed a riderless mount, and was seated again when we got the order to retreat on the column. We left about half our own cavalry behind in that charge, some dead, others wounded fatally, I hope. It was a case of *"sauve qui peut."* It was not until we were back with the column that it was possible to get into proper order. Our ranks were sadly depleted. Captain Dupont had a nasty scalp wound, but he was still on his mount, roaring out orders and cursing the Commandant of the column for this senseless fighting in the open. To make matters worse, it was decided that the main body of the column should push on, leaving behind the rearguard to follow as best they could!

In other words, the rearguard, which by now comprised about fifty exhausted men of our own Squadron, and about half that number of horses, and two depleted companies of Legion infantry, might go to ... The column had had enough of us apparently. Captain Dupont, with a bandaged head that was obviously giving him intense pain, went around like a madman. The two company commandants were pretty much the same, encouraging their men with kicks and curses. My leg was so stiff and painful that I could hardly put it to the ground, though I was compelled to do so for quite long stretches, for McCann was now sharing my mount.

Thus we hobbled along for the whole of that day, fighting and retreating, fighting and retreating. At night we kept up the struggle, digging ourselves in and working like machines rather than men.

The only thing that kept us going, the only thought that stimulated this super-human endurance, was the thought of getting into the hands of the Riffs. . . ,

We arrived at Fez the following morning, ninety men all told, the bedraggled remnants of one squadron of cavalry and two companies of infantry, who had fought continuously for two days against odds of about 300 to one. We were given a great welcome as we limped into Fez for what it was worth! Most of us went into hospital.

McCann and I, through the blunder of an orderly, were taken into an officers' ward, and should have been promptly moved out but for the intervention of Captain Dupont, himself a patient. *"Non!"* he roared. "If it is good enough for me, it is good enough for them. They have earned a decent bed. Let them stay! Besides, they are luck-bringers. They have as many fives as a cat!"

That was one of the few occasions on which Captain Dupont displayed any human feeling at all. McCann and I were in clover. We actually stayed between clean white sheets for a week!

Nothing like that had ever happened to us in the Legion before. It was heavenly! We had plenty of drinkable water, fresh fruit, and good rations. McCann grumbled, of course, because we were allowed only one pint of wine per day. By the third day he was itching to get back to duty, so that he might beg, borrow, buy or steal as much liquor as he could hold. Although our companionship endured for several years, I was never able to gauge the extent of Mac's capacity in this connection.

When we were reported fit and sent back to the Fifth Squadron the Skipper had gone before us we found big changes. It was now at full strength, and we scarcely knew half a dozen of the men. Most of the old crowd had either passed out, or were in hospital. I fancy our Commandant was glad to see us back among all those new faces:

"Ho! Ho! You crazy bandits! Back for more murder, yes? " *"Non, mon Capitaine"* answered McCann. "Back for more liquor!"

"Well, here are twenty francs. Drink at my expense. But don't get drunk, you *sales cochons !"*

As we were leaving, he called us back again.

"Barrington! You will take up your old position with me."

"Merci, mon Capitaine."

"Yelia! Yelia! Yelia!"

And we did! At top speed!

Feeling like new men, clean, spruce, fit, and rigged out with fresh uniforms, we went through the town looking for the dirtiest place we could find which was the Pit. We had money, and we had heard good news which we intended to celebrate. The Squadron was moving into Mequinez next day for three weeks' stay, and it was rumored that we should there join reinforcements for the Syrian Front. It sounded too good to be true. To leave Morocco for Syria meant a sea voyage and all the fun of a sea voyage, concerts, sports, carousals.

When we reached the Pit that night we found our comrades full of the news. They were arguing that if large numbers of the Legion were mustering at Mequinez, it could only be for the purpose of leaving Morocco for Syria. The general feeling seemed to be that such a transfer would be something in the nature of a picnic, compared to our experiences in Morocco. How little we knew!

Things were very lively in the Pit that night. A fresh bunch of girls had been imported from heaven knows where, and there was keen competition among the assembled legionnaires for some of them. One in particular, a black-eyed, impudent witch, who rejoiced in the name of Marcelle, was causing several heads to swim.

"Gee, boy! Can you beat this little girl?"

She was seated on McCann's knee, and the American was getting some nasty looks. He was also becoming well primed with liquor, so he did not care if it rained. They were men of the Fifth Squadron, too; but we did not know them very well as yet. One of them called out something in dirty French to Mac. He promptly picked up his glass and threw it at the offender, smashing it to splinters on the fellow's head.

Then the fun began. McCann literally threw the girl into my lap as his rival lunged forward, knife in hand. In another moment the two were engaged in a hand-to-hand fight in the center of the floor. McCann had the advantage of the longer reach. Already he had scraped the other's cheek, causing the blood to flow. Meanwhile, we remained seated at our tables and watched. Marcelle, perfidious devil, had placed her arms round my neck and was gently rubbing her soft cheek against mine.

"You come with me, *mon legionnaire,* yes? I show you."

She got down and began tugging at my tunic. The fight was now working up to fever heat, though it was clear that McCann was having it all his own way. It would be a lark, I thought, if McCann won the bout only to find that his own pal had gone off with his light-o'-love! The idea tickled me immensely . . . and off we went.

The girl took my hand, conducting me along what seemed like endless dark passages far from the noise and smell of the Pit. Presently we entered a small room. The place was quite sumptuously furnished. A lamp standing on a low octagonal table shed a soft glow about the colored rugs that adorned both walls and floor. There was a broad divan covered with gaily-colored cushions. It looked inviting and comfortable. We sat down. Marcelle held out her hand, and I covered it with francs my sign of good faith.

We lounged luxuriously on the softly-yielding cushions, and began to talk. This little girl of France was mightily interesting. Admittedly, she was but one of many such women who occupy the leisure hours of the Legion. Yet she was different from the general crowd. She had a wonderful command of languages. Her English was unusually good for one in such surroundings. She left Paris with a dancing troupe for Algiers.

"But there was so little money to send home, *mon legionnaire,* and there were always the men. So I came to Fez, where there are thousands of francs to be made/'

"But . . . d'you mean you wanted to come to Fez?" She raised her blackened eyebrows in astonishment.

"But, of course, *mon legionnaire!* There is much money here."

"They would not let you return to France?"

"They?"

"The men who brought you out to Africa."

"But, yes. I could go back to Paris, to Le Mans—but I did not wish."

"Why not?"

"Ah, *mon legionnaire,* my mother and father are so old. They have little money, and there are many brothers and sisters. It was terrible . . . terrible! My father go to work when he cannot walk. And for me, it was always the dancing . . . dancing, dancing, dancing. Very hard work and so little money. Here, in Fez, I get much money, and I can send it to my people. They are happy now."

"Do they know?"

"Know?"

"Yes. About . . . this."

"Non! Non! Non! But how could they? They think I am dancing, dancing all the time."

"And do you like . . . this?"

She laughed, shaking the cushions.

"But *monsieur* is too funny!"

And she fairly rocked with laughter.

"But you will go back to France . . . one day?"

She looked at me with a quizzical smile, her full red lips thrust forward, and slowly shook her head.

"Who can say? Does it matter? Perhaps I find rich man, who will marry me. You are very rich, in England, yes? "

"Very rich I don't think!"

“Very rich, but you do not think about it, yes? You not care much for money, yes? ”

“That’s right.”

“Oh! *Monsieur!* What is that?”

There came a sound of unsteady and very heavy feet in the passage without. A voice was bawling my name, and that voice was thoroughly American.

“Hey! Barrington! Where in hell are you! Barrington, you damned limejuicer!”

I left the room very hurriedly and ran down the passage towards the bawling voice. By the sound of things, I guessed that McCann was well loaded, and was attempting a rough house. When I came upon him he was struggling with the fat Greek and three or four men of the Legion. They appeared to be having a very lively time with him.

When he saw me, however, he quietened down. We stared at each other for a moment. He presented a truly wonderful picture, for his tunic was tom open, his mouth was bleeding, and one ear was gashed and swollen to the size of two.

“Hey, Barrington! What’ja think you are the ‘sheek’ of Araby!” “I’ve been looking for you all over the place,” I said, taking his arm and leading him along.

“Yeah? Been having a nice Sunday afternoon walk, eh? You big stiff! Where’s that bird?”

“What bird? Marcelle? ”

"Shoor!”

"She cleared off. She doesn’t like soldiers with knives.”

"Nope! Whadda she want, anyway? Bitch! Right, boy, let’s go. Let’s find some hooch, eh?”

"Just one, and then back to barracks.”

"Shoor! Back to barracks like good little soldiers. . . . But that bird, Barrington, Marcelle what's-her-name. . . . Nice little cup o' tea, eh?"

But I don't think McCann ever really forgave me for that night's work. In spite of the load he took home, he appeared to have a lot of the details quite correct next morning.

"1 guess you're a wide 'un, Barrington." And

he said it with a queer sort of look.

"I told you, Mac. All the United States wouldn't have held her back to watch that grisly fight of yours."

"I left her on your knee, you goldarned limejuicer!"

"She slipped off."

"And so did you?"

"Isn't it a bit futile, Mac, this argument? You never know what we'd find in Mequinez."

"No, boy, and it won't be parked in *your* arms, I guess!"

We were in the horse lines, grooming, when this illuminating conversation was going on. I was bending to the job when I became conscious of an approaching shadow. I looked up. It was Captain Dupont, and with him were two strangers, two elderly people, a man and a woman, respectable, well-dressed, entirely out of place just there. I stood at attention, staring. This whitehaired old couple, obviously well-to-do, were French. I could see that.

"Barrington," said Captain Dupont, looking me straight in the eye, as if he were warning me, "Madame and Monsieur Beaumont! They have journeyed from France to speak of their son, Lieutenant Beaumont, who died so gallantly."

I turned to the old couple and saluted smartly, then waited. McCann stood by, also waiting.

"You are English, Legionnaire Barrington?"

"All oui, Madame."

“And you, Legionnaire Mek . . . Mekkan?”

“American, Madame.”

“Le Commandant tells me you were by my son, Lieutenant Beaumont, when he fell in the retreat from Bab Mezab . . . tell me . . .”

It was one of the worst experiences that McCann and I encountered during the whole of our time in the Legion. There was the little, silver-haired old lady, calm but terribly pathetic, and her husband, a tall, well set-up fellow, with iron-grey hair and trim beard, spruce, speckless, and obviously at the mercy of the determined little woman. Apparently she had dragged him all that way from Brittany to Fez merely to hear the details at first hand of the bravery of her gallant son who had died fighting after so short a service. Before she had got half-way through with her fire of questions, McCann and I were perspiring freely. Captain Dupont stood aside, watching us, listening. A slip on the part of either of us, and we should have had a dose of cells as soon as the old couple had departed. Luckily for us, we passed through the ordeal with flying colors. The old couple were intensely grateful, immensely relieved to hear our version of the story of his end. They bowed, thanking us with tears in their eyes, and went.

“Aren’t we a pair of prize liars, Mac?” I remarked, when they had got out of earshot.

“D’ja ever see anything like it in your life, boy? But, say! Has somebody been spilling the beans? D’ja think the lil’ old dame had some idea of the truth?”

“Well, there might be something in that, Mac. Otherwise, why come out all this way, and to a god-forsaken place like Fez, merely to hear how their son died, from two common legionnaires who happened to be alongside him at the time?"

“Shoor, boy. I reckon some fool guy has said a bibful, and the old folks got wind of it. That’s the only explanation I can think of. D’ja get me?"

"Dirty swine, whoever he was!"

"You've said it, boy. The son of a bitch ought to be . . ."

And McCann said what he thought should be done with such a person.

At all events, the strange episode had tried us both so thoroughly that we quite forgot our differences of the previous evening, and were good and regular fellers once more. In the afternoon of that same day we moved out of Fez to join a big column at Mequinez.

Moving with a column was always a popular event with the boys of the Legion, not so much because of the big muster and the varied life of such a large body of men on the march, but because of the women.

It should be stated, as descriptive of the life in those regions, that a properly organized party of women always travelled with a column on the march, and they formed a definite quarter of the camp known as the *Maison de Tolerance.* For a column numbering two thousand men there would be some fifty such women. Truly a hard life for the little band. It is conceivable that the little Marcelle would have preferred her dancing, always dancing, to such hardship. Of course, the officers were lords of creation in matters of this kind. Other ranks might expect to spend ten to twenty francs if the column was long enough on the march.

It was the existence of this quarter of the camp that got four of us out of a very tight comer during our second week in Mequinez. McCann and I had become acquainted with two other legionnaires named Cheyes and Scora, both of whom spoke very decent English. Cheyes had been a Belgian banker who had learned his English as a refugee in England during the Great War. He had been employed by the G.P.O. until the end of 1918, and had then returned to his own country. He had settled down in the bank for a number of years, and then had come the temptation to mix his master's money with his own. He escaped to France, and when that country became too hot to hold him, he joined the Legion.

Scora was an ex-officer of the Italian Army. In a passionate affair with his wife's Jover he had committed murder. Otherwise, I must confess, he was a very decent fellow, staunch, dependable, and ready for any sort of lark or adventure.

One night the four of us decided to raid the supply hut, so that we might taste the good things of life for once. Our big difficulty was to get past the guard of mounted Senegalese who were posted for duty there; but as, by this time, we had learned a few Arabian tricks regarding this delicate job of getting by a sentry, we got inside the hut without mishap.

We each had a sack and worked by shaded matchlight, which was something of a task. We should have been all right had McCann and Cheyes not found the good Jamaica rum. It was raw stuff, and the pair of them had a rather liberal dose. In finding our way out, with the loaded sacks on our backs, McCann stumbled over a case, and knocked up against Scora who was carrying a light. The result was that Scora dropped his light on to some straw and other packing material and in a flash the stuff was blazing furiously. We made a feeble attempt to stamp it out, but already the alarm had been given and there was nothing for it but to get clear in the general confusion.

No sooner had we got clear of the place than it became a mass of flames shooting skywards. The whole camp was roused. Whistles were blowing, officers were cursing and shouting, horses were stampeding, and men were racing along with buckets of water. As McCann expressed it, hell broke loose.

We dumped our precious sacks as quickly as we could, and ran towards our own lines. Captain Dupont was bawling our names and cursing us for sons of dogs and pigs. When he saw us he began to use his boot in real earnest, asking us where we had been, why we were not in the lines when we heard the order to muster.

McCann, with a genius for that sort of thing, rose to the occasion marvelously. He informed the Skipper that when we heard the commotion we were at the *Maison de Tolerance!* And we had returned to our own quarters at the double. Fortunately for us, Captain Dupont accepted the explanation, merely contenting himself with a few sanguinary epithets descriptive of our utter worthlessness in the Legion or anywhere else.

We were at once dispatched to join the fire fighters. And what a sight it was! The fire was well under by now, but there was such a crowd of troops round the place, such confusion and turmoil, that looting seemed to be the order of the day. There was a regular raid on the officers' rations, and the officers had a pretty shrewd idea as to what was happening, but since the whole camp was thoroughly out of hand they could do nothing but curse and gnash their teeth.

For days afterwards we ate real white bread plentifully smeared with butter stuff we had not tasted for years! And as for the liquor that was lost, men stood on parade roaring drunk for the next few days, McCann not the least of these!

Had the origin of that mysterious fire ever been discovered, four English-speaking legionnaires would have been given a couple of years in chains, and, greater calamity still, as we thought at the time, those same four would have missed the draft for Syria.

We later came to wish that we had been found out, for a few years in chains could not have been worse than a few years on the Syrian Front!

CHAPTER VIII: SIEGE OF DAMASCUS

We went on that happy sea voyage totally unaware of the fact that we should be just in time for the revolution in Syria. We thought we knew all about native barbarities and cruelties. We had yet to meet the Jebel Druses, perhaps the most ancient, and certainly the most ferocious race on earth.

Who and what are the Druses? Nobody knows. They are an extraordinary people, who show allegiance to no other power. Their kingdom is in the north of Syria, comprising the whole of the southern range of Mount Lebanon and the western slope of Anti-Lebanon. Nearly half of the Druse nation inhabit the Jebel Druse, a mountainous district to the southwest of Damascus. In the days of Turkish rule they were constantly at war with the Turks, refusing to recognize either taxation or conscription; and since the Protectorate came under French control their attitude has remained unaltered.

According to their own tradition, they were originally Arab tribes from Yemen, who migrated to Mesopotamia, thence to the neighborhood of Aleppo, and later, further south. Another tradition connects them with China. Still another traces their origin to the Cuthites, or Kurds, with whom, after the second captivity of Israel, Esarhaddon re-peopled the wasted strongholds of Samaria. More than a thousand years later the Mardi, a warlike tribe of Persian extraction, were sent into the country by Constantine IV., to act as a bulwark against the Mohammedan invasion. The Arabs also, in sweeping through the mountain fastnesses, left a permanent impression upon them. So, too, did their association with the Turks.

Consider, then, this hybrid mixture Cuthites, Mardi, Arabs, and in all probability, Chinese, a regular conglomeration of various races of Mohammedans, and the very pick of the fighting tribes of those races, and you have the combination that forms one of the strangest of nationalities to-day the modern Druses. What they do not know about the horrors of war, considering their history and associations, is not knowable. There is no other race on earth, I am convinced, that can be so ferocious, so utterly fearless, and so callously brutal in the interests of their racial principles and religion.

When, therefore, you have such a race at war with the equally ferocious and brutal French Foreign Legion what sort of situation is likely to ensue?

The Syrian revolution into which we so unwarily stumbled had its origin in Sueida, the very heart of the Jebel Druse's country. A French Governor had been sent there, with a detachment of 600 troops, whose chief mission was to maintain peace among the Druses. But it appeared he did not fully appreciate the delicacy of his mission. There were complaints of harsh and unjust treatment. The independent spirit of the Druses revolted against the idea of being under the heel of anyone. For years they had fought the Turks. They had succeeded largely in obtaining what they considered their just rights, and they held and guarded them most zealously. They were not to be dictated to by the French.

One memorable night the Druses rose in massed numbers and massacred every man in the Sueida Citadel, including the Governor himself. Down on the plains the Bedouins took advantage of the chaotic situation. As the French were unprepared, having neither sufficient troops nor guns, looting and pillaging were rife. The Druses were at war with the French, the Bedouin Christians fought the Mohammedans and the

Druses, whom they hated, the Mohammedans fought the Druses, the Bedouins and the French! A more bloodthirsty and chaotic situation could not be imagined.

At first, Damascus was peaceful, and it is probable that the Druses' quarter would have remained quiet, had it not been for the high-handedness and generally insulting behavior of the French troops especially in regard to the Druse and Mohammedan women who wore the veil. The result was they rose like one man, and started to fight French, Bedouin and Mohammedan. The General commanding the troops in Damascus clamored for reinforcements. And so we were rushed from Beirut to Damascus.

Our first glimpse of that beautiful and highly-colored city was somewhat hurried and rather hectic, since we were engaged storming the North Gate, with the purpose of fighting our way through the streets to the assistance of the French troops and the Senegalese, who were more or less imprisoned in the Citadel. The city was in a state of siege. We did not know who was enemy and who friend. The beautiful gardens gained our attention only because they held snipers.

At the North Gate Captain Dupont decided to send a party of a dozen men to reconnoiter. McCann and I, and our two friends, Scora and Cheyes, were included in the patrol. We went along the streets at a walk-march. It was all too quiet, too sinister, for so gay and colorful a city. The narrow and badly paved streets were thick with dust. On either side of them, where are the arched niches that do duty for shops, there was little to be seen, for the old merchants had shuttered and barricaded their stalls.

From the distance came the sounds of firing, the rapid shooting of Mausers and machine guns. We proceeded cautiously, feeling that at any moment fire would break out upon us. The sergeant who led us was making for the "Street called straight," as being the extent of his reconnoiter. We were rounding a corner into this historic thoroughfare when a volley of bullets greeted us. Some of our men pitched from their saddles, horses screamed, but, along with others, I managed to turn about and race towards the shelter of a building.

There we turned and faced the mob. It was our first taste of the Jebel Druses, and it looked as if half their forces were concentrated on that corner. They descended upon us, one mad rush, yelling and foaming at the mouth, screaming "Down with the infidels! Drive the French to the sea!" They certainly drove us back. There was nothing for it but to gallop back the way we had come. We went along the streets as if chased by ten thousand devils, the bullets flying right and left.

Of the twelve men who formed that patrol, only three of us reached the North Gate again McCann, Scora and myself! What happened to the others heaven alone knew. We reported to our Commandant that hundreds of Druses were charging down towards the Gates, and that we had lost the remainder of the patrol. Quickly he gave the order to mount and line up for the charge against the oncoming mob. It mattered not to Captain Dupont how many Druses there were. His orders were to relieve the Citadel, and he would take the gallant Fifth Squadron there, whatever the cost.

With saber in one hand and Mauser pistol in the other, guiding the horses with our knees, we charged towards the oncoming hordes. Never shall I forget those screaming Druses. The noise they made was deafening. They were answered in a score of languages by the legionnaires. I found myself in a hand-to-hand fight with a big Druse, and my career would have ended then but for a timely shot from McCann's pistol.

Nevertheless, they were too many for us. They drove us back with a loss of nearly a hundred men—dead and wounded left in the streets at the mercy of the Druses' women. We were back at the Gates, but only for a few moments. The Legion is never permitted to take "No" for an answer. We had hardly got ourselves into shape again, when Captain Dupont, thirsting for blood, gave the order to charge.

Once again we rushed towards that bunch of screaming maniacs, the Skipper yelling at the top of his voice *"Vive la Legion! Vive la Legion! Sabre un Min! Sabre un Min! "* ordering his dirty pigs and dogs to charge and fight like the Legion! We did, cutting in, slashing and firing like so many madmen. And this time we succeeded in cutting our way through the Druses massed in the narrow streets, right up to the Citadel itself, where we were joined by that famous fighting cavalry, the Spahis Moroccan. Never were we more thankful to see that mob. We entered the Citadel with them, and joined the rest of the French troops. I had the temerity to ask Captain Dupont what we were to do about the wounded left in the streets. He turned to me, his eyes still wild, blood trickling down his cheek.

"To hell with the wounded!" he cried. "The Druses have them, and by this time are tearing them limb from limb. Barrington, *nom du chien,* when you know the Druses as I do you won't ask such fool questions! "

I turned away, feeling a bit sick. The Druses, then, were much of a pattern with the Riffs, only more so. And we had heard such wonderful stories about the beautiful city of Damascus! We were lined up for roll call. Of the original 250 we had lost 130. It was a big loss, in all conscience, but we must have inflicted greater losses on the Druses.

Orders were then given for us to unsaddle and attend to the horses. We found plenty of good stabling in the Citadel. In fact, there appeared to be plenty of everything in that massive fortress. The courtyard alone would hold five thousand men. Also, there were sufficient supplies and water there to sustain a garrison for eighteen months. It was a town in itself. G

The Citadel of Damascus stands within the western walls. It is a huge quadrangular structure, about 350 yards long and 300 in width, with projecting towers and, at one time, was surrounded by a moat, which had become a good deal dilapidated at the time of the siege by the Druses.

After our stormy entry we were all pretty well exhausted, and once we had seen to our mounts we were glad enough to get down to rest. McCann and I lay on our saddlebags and were soon fast asleep. We had not been in that position long, however, before we were roused by a fusillade of shots, and the cries of sentries:

"Aux arme ! Aux armes! Aux armes! "

We jumped to our feet and fell in with the squadron. It appeared that the Druses were far from being subdued. Street fighting had broken out, and there were orders for large patrols. Four to five hundred men were turned out of the Citadel that night, leaving only a skeleton garrison, and patrolled the streets, killing and slaying on sight. All night long it went on. Men and women were shot and stabbed merely because they were abroad at all. On the slightest pretext, the troops would enter a house, loot the goods, manhandle the women, and kill the men. Before dawn, the whole city was thoroughly subdued.

I have seen few more ghastly spectacles. The famous gardens were splashed with blood, and so were the marble basins of the fountains. And this in Damascus the city famous for its lovely fountains and cooing doves!

Throughout the following day the bloodthirsty crew were allowed to sleep off their debauchery and butchery. When night came again, the city was in boundary. Troops went about to the drinking shops and cafes, mixing with the populace, and generally enjoying themselves! It was the kind of transformation that could not possibly happen elsewhere.

For the next two or three days all seemed fairly quiet. Our own men, of the Fifth Squadron, had settled in. Many of them had been out exploring the city, finding all sorts of places, some worth recommending, and some quite otherwise. I remember one incident particularly.

Martial law had been declared, which meant, among other things, that everyone must be indoors by ten p.m. One night a Spanish boy belonging to our Squadron was posted missing. His name was Delius, and he was by way of being a fresher. He had been swaggering to Scora, McCann and me, telling us in his hotheaded fashion what he intended to do with the first veiled woman he found wandering alone. We were all drinking in a little place we had found and had made our unofficial headquarters, which rejoiced in the name of the *Cafe Noir.* At the time we paid little attention to his bragging. Such drunken talk was common enough.

It was only after he had been posted missing that we remembered what he had said. I talked the matter over with Scora and Mac, and it was decided to report to Captain Dupont. The Skipper flew into a rage when I awakened him, at somewhere past midnight, to tell him the story. Nevertheless, he turned out a patrol and headed it himself. My two chums and I were naturally included in the patrol, which was made up of twelve legionnaires, a sergeant and Captain Dupont.

We turned out about one in the morning. The streets were deserted, of course. We concentrated on all the streets and buildings in the vicinity of the *Cafe Noir,* though the fellow might have wandered to the other side of the city. Certainly, he was mad enough for anything. We did not come upon a soul, save the patrols of other companies, and to these we gave information about the missing legionnaire. Round and round we went, making a more extended area as we searched. We kept it up for about three hours before the mystery was solved, and by Captain Dupont himself. He was marching a little ahead and to the left of us, keeping close to the buildings, when he suddenly halted and gave vent to his favorite expression:

"Sacre bon Dieu ! "

He picked up something from behind a garbage can, picked it up as if it were tainted, on the tip of his saber. It was a legionnaire's cap. We looked up at the ancient dwelling nearby. The front of it was in darkness. The Skipper roared out an order and we went forward and rapped on the great heavy door with our pistols, calling on the occupants to open it. Presently an old Druse appeared. He was questioned concerning the cap, which was obviously that of the missing Delius.

The grey-headed and grey-bearded old patriarch threw up his hands and swore before Allah that he knew nothing of the missing soldier.

There was no one in his abode except the members of his own family. He protested that he was a peace-loving citizen of Damascus, that he desired nothing better than the protection of the French Republic, that he loved the French as a brother, mother and father. . . .

The sergeant looked at the Skipper. The latter nodded, ever so slightly. The next minute the old man found himself on his head in the middle of the street. We all trooped through the doorway, across a courtyard, and into the house. In a room at the back of the building we came upon the poor devil, Delius. He would not brag any more. His bragging days were over. But, God! What a sight he was! The women who had done this thing were caught in the very act, dripping knives in their hands. There were three of them, crouching by the wall, veiled women in voluminous black clothing. Except for that first scream when we broke into the room, none of them made a sound. They just crouched there, knowing that this was the end for them.

We stared at that ghastly naked thing that had been their plaything, stretched out across the middle of the floor, spreadeagled, feet, hands and neck secured to the floor boards. The face of Delius was no longer a face, but a bestial, shocking, nauseating travesty of a face; for those women had been up to their old tricks.

When the enormity of this horrible thing that had been done to our comrade impressed itself upon us, we turned suddenly mad. A dozen pairs of hands fell upon those women, tearing away the veils and clothing too. They would have been clawed to death there and then, had it not been for the Skipper. He had ways of his own. He could mete out punishment as subtle and diabolical as any native of the East. . . .

Delius was still breathing as we loosened the thongs that bound him to the floor, still warm as we re-adjusted him in the semblance of a man. While we attended to him, others searched the building, and presently returned with two more women. Now they were five, all of them incredibly young for such an act. Lined up against the wall, they stared at us, their unveiled features working with terror. One by one they were divested of clothing and slung up to the beams of the ceiling, feet first; and no army could have stopped those enraged soldiers, fiendishly exulting in the punishing of the women, two of whom were little more than girls, for this lascivious mutilation of a comrade. The Skipper gave no order to fire, knowing that the mad patrol would fire . . . again and again . . . according to the dictates of their insane impulses and fancies. . . . The screams of those women filled the night, or rather the early dawn, renting the air in terrific crescendo . . . piercing enough to awaken even the dead Delius.

Thus it is that I can never hear anyone mouth that silly threat "You'll be shot at dawn! " without picturing once again that terrifying scene, those five women . . . screaming . . . streaming.

That morning McCann made a vow that he would never touch another woman as long as he remained in Syria. And he broke his vow on the evening of that same day! I had noticed for two or three days past that he had become very intimate with a German girl who served at the bar of the *Cafe Noir.* Elspeth was a comely sort of wench, generous as to curves, a trifle blousy, perhaps, and hardly to my taste, though something of a prize in that environment. McCann, to use one of his own expressions, had properly fallen for her.

That night, when we were endeavoring to forget the picture of Delius by guzzling gallons of wine, McCann by his attitude began to make it clear that Elspeth was his property, and if any fellow legionnaire liked to dispute his right he was at liberty to do so. He deserted the table at which Scora and I were seated and spent most of the night leaning on the bar talking to his Elspeth. I do not suggest that the hard-bitten, tough, man-eating fellow was in love. But he came as near to that perilous state as any man of his make-up possibly could. In short, he was just crazy about Elspeth!

It was too comical for words. It was almost unbelievable that a chap of McCann's kidney should show his hand before everybody in that fashion. It was not an ordinary case of picking the girl up and carrying her off, and fighting any man who attempted to interfere with him. There was something deeper in it than that. Which was ridiculous, of course, in such a place. No decent girl would have been found dead in that sink of iniquity. And yet, there was Mac, of all men, treating her as if she were some homely girl with whom he might, did he behave himself, go out one Sunday afternoon!

It was funnier still when we had returned to the Citadel and had started to kip down for the night. McCann began to talk about Elspeth, and he continued to talk about Elspeth, to enlarge upon her charms, her little mannerisms, her beauty far into the night. And I chuckled and chuckled. It was as good as a comedy. My heavens, I thought, this is McCann, the fellow with the India rubber jowl and the India-rubber soul, the crack shot of the Squadron, the biggest drinker and the biggest killer in the whole Legion cavalry and he's smitten! The great, big, long-legged sap!

"Oh, yeah?" I said, feeling a bit fed, "I guess she is all that.

What about it? She's well, just a nice cup o' tea!"

And in a minute he was sprawling over me, his great hands fastening round my throat. I spluttered and gasped and kicked, and he wouldn't let go until I landed him a beauty in the belly. "You damned fool!" I shouted, "what's bitten you! Are you going *cafard* jumping on a fellow like that?"

"Cafard 1" he snorted. "Whaddy mean, nice cup o' tea! You dirty limejuicer! Get this! She's O.K. Keep off! D'ja hear? I guess she's different."

I sat up and stared at him in amazement. He had himself referred to hundreds of girls as a "nice cup o' tea," since we rarely met any other sort of girl. Was he imagining he had met one now? Heavens! This was no comedy. It was deadly serious.

He calmed down and began to talk rationally, at least, as rationally as any man might while suffering such a delusion. "I'm going to have a shot at deserting, Barrington. D'ja get that, boy? Over the border to Palestine and freedom, and little Elspeth will be there!"

"Good God, Mac, you must be crazy!"

Little Elspeth, indeed! I considered it high time for me to start in with the talking. There had been quite enough from his end. So, I talked and talked, for hours it seemed. I referred to our long comradeship together, to the various little things we had been through together. The Legion was going to be more than hell on earth for me if he persisted in this insane idea of attempting desertion for the sake of *little* Elspeth. If it was a case of getting over the border, why shouldn't we have a shot at it together. We had often considered the possibility, even in Morocco. The chances were better in Syria, for the border that led to Palestine was British, and therefore . . .

"Shoor, boy! You join the wagon! Let's go! Once we're over the

border, she'll be useful." "She?" I said doubtfully.

"Shoor! Elspeth! Why, yes?"

But I didn't see it that way, and I said as much. A woman would be an encumbrance. We should never do it with skirts at our heels. Heavens! Wasn't the job difficult enough? But I saw it was no good arguing with McCann in his present state of mind. So, I just fell asleep on it.

On the afternoon of the following day Mac persuaded me to go along to the *Cafe Noir* for the purpose of discussing the project with Elspeth. I cannot say that it appealed to me in that light. I had always visualized our escape as a well-planned, two-handed affair. Supposing we fell in with the Partisans, as we were almost certain to do. What should we do with a woman then. *What would the Partisans do with her and us ?*

I was still arguing when we reached the cafe. We found our way to the bar. Elspeth was nowhere to be seen. Mac made some inquiries and learned that she was entertaining a friend in her room. He winked at me.

"I guess she's got a girlfriend up there from one of the other dives. When you see her, boy, perhaps you'll change your mind about having a woman along. Kim on! Let's go! Let's dig 'em out 1 What say? "

We found our way to Elspeth's room and, by way of a surprise, charged straight in. It was a surprise all right, for both parties, but more especially for McCann. Elspeth, we saw, was lolling comfortably on the knees of a sergeant of the Spahis Tunisian.

With a roar like a wild bull, McCann leapt at him, knocking the girl across the room in his mad rush, and taking the sergeant's head in his stride. Down the two of them went, with Mac uppermost, pounding the head of the unfortunate sergeant unmercifully. Even after the fellow had lost consciousness I had the greatest difficulty in dragging the American off. When, finally, he did rise to his feet, he just looked at the girl, who was reclining serenely on a couch.

Then he began to tell her things. As he would put it, he spilled a bibful. His choice of oaths was uncommonly clever for one who spoke under great stress. The girl paled under his terrific lash of epithets. He summed up her character, from her birth on a dunghill to her inevitable demise in a brothel, using an amazing variety of adjectives in the process. When he was through his mouth foamed and the sweat was pouring down his leathery face. He could not have been more exhausted after a ten hour stretch with the guns.

"But," I remarked, "what could you expect in a place like that?"

Mac shook his head. We were retracing our steps towards the Citadel. Then, at last, it came with a rush.

"Gosh, boy! Can you beat it! It's still possible for the bitches to kid the old fool, McCann!"

It certainly was a big blow to his self-respect. And it takes a man a long time to get over that, especially a man of the McCann type!

CHAPTER IX: THE MESME BLUNDER

We left Damascus for Mesme, a small outpost on the railway that consisted of a couple of sheds and a well, on another of those futile missions which the French love to perpetrate on a long-suffering Legion. We started out at dawn, 130 legionnaires heading a strong column of Senegalese and Spahis Moroccan. The streets were ominously quiet as we passed through the South Gates for the plains. Not a single shot greeted us, though we knew as we proceeded along the open country that it was infested with Druses and Bedouins. Evidently they were satisfied to permit us to go peacefully into the death-trap, for that was what it turned out to be.

A bunch of Partisans served as our guides. As we marched, the sun came forth in all his glory. It was difficult to believe that a revolutionary war was raging up and down this vast and beautiful country of Syria. We crossed that huge expanse of green sward which stretches from the base of Jebel Kasyun mountain right over the plains to the city. Hereabout are dotted scores of villages and gardens, in which all the trees of forest and field blend their many shades and colors in an effort to keep the desert at bay.

Then came the great rolling sand dunes, shimmering in the sun, and behind them lurked death or worse. By midday the heat was intense, horses and men were perspiring freely. We followed the railway track for some six hours and then came suddenly upon the outpost of Mesme, a desolate and tiny encampment on the rail embankment some few hundred yards ahead of us.

I wondered if the detachment guarding this God-forsaken spot could see us. I wondered what were in their thoughts. I was soon to know. Captain Dupont sent a sergeant and myself forward to see that all was well. We approached at a trot and were met by a French cavalry officer, who asked us if we were his relief.

<r Thank God! " He exclaimed. " We lost fifty more men last night! "

A cheerful welcome, to be sure! The whole column came up and we of the thinned-out Fifth proceeded to take over. The bedraggled detachment turned out. What a sight they were! One half of them were either sick or wounded. A number were slung between mules and went back as stretcher cases. They set off, escorted by the column that had brought us out, leaving us behind to defend what had obviously proved to be a deadly little outpost. And yet, nothing happened for nearly a week! During that time we were busy building fortifications. We covered a big area with barbed wire and wide water trenches, completely closing in the well and the two buildings that served as a railway station. We also built an elaborate system of trenches and dug-outs. We were leaving nothing to chance!

One night, while I was on guard, we had the first indication of the enemy's nearness. I was on number i post and twenty paces away was number 2. This was occupied by a little Russian whom we knew as Broski. In patrolling my post, and he his, we met at frequent intervals, and I soon found that he was a regular bundle of nerves. He remarked more than once that he felt something was going to happen. I told him not to be a fool. The place was getting on my nerves, too, but I was not going to admit it. We had had a week of it, and nothing had happened. We knew, almost to a man, that this interval of peace was just a way the Druses had of playing with their enemies. Their subtle game was to give us a sense of security and peace, and then make their sudden descent.

Occasionally I looked towards Broski's post. We had built our fortifications on a quadrangular plan, and the extent of his post was to take in an angle of the trenches, meet number 3, and about turn, so that for a few seconds now and again, I could not see him at all. Once I heard, or thought I heard something like a gurgle, but as our relief was then due, I took no notice.

A corporal came along with my relief. I reported all well, and fell in with the guard. When we reached Broski's post we found him crumpled up on the ground, decapitated! It was a horrible sight.

The subtlety and swiftness of it! I had seen and spoken to him only a few moments before, and he was never more than twenty paces from me! I explained all this, and also that, only a few moments before the relief came I had imagined I'd heard a slight gurgling sound.

The whole camp was roused. I repeated my story to Captain Dupont.

"Sa ere bon Dieu ! But the devils are going to attack. This is the sign! "

It was impossible to double the guard. We had not sufficient men. The only way was to put everybody on guard, and for the remainder of the night every man stood to.

As dawn broke there appeared from all sides a huge force of massed Druses. We opened fire with machine guns, bringing down their front ranks. On they came, fiendishly yelling their battle cry:

"Yelia! Jo Allah! Yelia!"

As we sprayed them with the machine guns their ranks thinned, but others quickly took their places and they continued the charge, apparently with the intention of taking us in one frantic rush. They reached our fortifications and many of them leapt across the barbed wire trenches, coming to a hand-to-hand fighting with their long, old-fashioned swords. Somehow we managed to drive them off, and as soon as that happened we had the machine guns going again. Gradually they were forced back and back. They had lost their big opportunity. Had they carried on for a few minutes longer after clearing the water trenches we should surely have had to give up.

They retreated across the plains, leaving their dead and wounded scattered on the ground. Captain Dupont decided to make a sortie. We mounted fifty men, and, with the Skipper at the head, dashed off after the stragglers. But we hardly got away from the outpost when we had the shock of our lives. We had been led into a trap. The dead and wounded Druses who had lain scattered about the plains, or, at all events, a vast number of them, suddenly rose like one man, very much alive, and surrounded us. In a moment the whole bunch of us were thrown into the greatest confusion. Our

Commandant was cursing and shouting the order to "*Sabre un min 1 Sabre un min! *'* The sound of his voice kept us more or less together. We turned and raced back towards the post, using saber and Mauser as we galloped along.

We lost eleven men in that sortie. The Druses did not follow us, being content to remain outside the range of the machine guns. There came a lull towards late afternoon. Volunteers were called to go out for our men lost in the sortie. McCann and I joined four Germans in this mad quest. We crept slowly along the ground, inch by inch, firing at every body of a Druse we came upon, lest he rise suddenly and trap us again.

McCann and I had picked up two of our comrades and were making our way back again, when a number of shots rang out, followed by terrific screams. We looked round to find two of the Germans held by a number of Druses, while some of their kinsmen proceeded to disembowel them as they stood on their feet. The other two Germans were racing towards us. We dropped our comrades and started to fire.

"You ruddy sons of bitches!" roared McCann, and got to his feet as if he would fight the bunch of them.

I deliberately tripped him and he came down heavily.

"Don't be a fool!" I shouted. "It's sheer suicide to tackle that bunch. Come on! Let's get back!"

Between us we dragged the two wounded along to the post, and reported to Captain Dupont. We had brought in two men, but only at the expense of having two others captured and tortured before our very eyes.

That night our new-found chum, Scora, the Italian, was found spread-eagled on the barbed wire, his sightless eyes staring up at the sky. He was alive, but had gone *cafard.* In fact, it was the noise of his gibbering that drew our attention to him. It was what the Skipper was pleased to call a sign, for the next morning fighting was renewed, and continued throughout the day. Altogether, we stood up to it for three days and three nights before the relief came along.

We were reinforced to our full strength. Three companies of the Legion infantry, with bombing 'planes, made a clean sweep of the area, and peace reigned once more.

It seems incredible to relate, but is nevertheless true, that a couple of days later we were moving about the area with perfect freedom, visiting neighboring villages, and generally enjoying our sojourn in the district as if there had never been such a thing as the Mesme Blunder. Actually, the revolution was still proceeding in other parts of the country, but since the Druses had withdrawn their forces, temporarily, no doubt, in order to reinforce other districts, we were left in peace.

Trains with troops and ammunitions were passing through our fortifications on their way to Rashaya, and we gave them all a hearty send-off, knowing we should follow as soon as the region grew hot enough!

Some three or four miles beyond Mesme we had found a native village, little more than an Arab encampment, which we visited frequently. For one thing, it provided a good terminus for our exercising ground, and for another, it boasted a wineshop of sorts. One shabby dwelling had a particular attraction for the Skipper. We never knew why, but we could guess. It became a practice for him to take a small detachment out there every morning, ostensibly for exercise, actually for the Commandant's pleasure, to say nothing of his officers. The little wineshop did a roaring trade, for a legionnaire must take his pleasures when and how he can.

McCann and I were riding leisurely through the village one morning when we came upon a most unusual incident, unusual, that is, for such an out-of-the-way village. Riding as leisurely towards us were three natives, their saddles, clothes and trappings simply shouted wealth. Involuntarily we pulled up in the narrow street, and waited for them to approach.

"Gosh! Barrington, boy, what have we got here?"

"Looks like the Sultan of all the Asias and his missus."

For we saw, as they came closer, that one of the riders was a veiled woman. She, like the two men, rode astride, and from beneath her voluminous black silk robes we caught a glimpse of slim, prettily-fashioned top boots and the breeches of scarlet satin stuff. We stared hard as they sauntered slowly by. We had not seen anything quite like this for a long time. It was a regular feast for our eyes. Even in that land of strange and unexpected things, this picture seemed totally out of keeping.

As they rode by the woman turned her head, staring at us, we knew, through the thick veil. The men ignored us and our looks of inquiry, riding haughtily, as only Eastern nobility knows how. They still looked to their front when the woman skewed round in her saddle and took another long look at us, displaying a wealth of satin-covered leg in the process.

Naturally we were vastly intrigued by her devil-may- care curiosity.

"Waal!" gasped McCann, "can you beat *that!*"

"A good tiling for her the Skipper isn't around, Mac?" "Shoorest thing you know, boy! He's busy. If he'd seen that bitch he'd have been after her . . . red breeches and all!"

"She must be very young . . . capering about like that?" "Yeah. A young enough kid, I reckon. Otherwise, she wouldn't have been so darned curious about a couple of legionnaires. The Skipper would have been all over her, even if it had meant starting another war. Gosh, boy! Let's go!"

McCann started to canter along the street, and I followed. But already the other three had spurred their mounts to a gallop, and soon had cleared the village for the open country. That was where we drew rein. You never can trust a native, no matter what his caste. It might have been a trap, it might not; but we were taking no risks in a neighborhood where anything might happen at any time. We turned and retraced our steps back to the Mesme post. For the remainder of that day we talked of little else but the mysterious veiled rider, and her impudent stares, her youthful curiosity.

We contrived to be in the same place the following day at about the same time; but nothing happened. We tried day after day, with the same result. The mysterious woman rider had vanished. We refrained from making any inquiries, since we did not wish others to know of what we had seen. “I expect she’s somewhere in Beirut by this time,” commented McCann, when, a week later, we were still searching. He was wrong, however. Her home was but a few miles from that village, as I was to learn very soon.

CHAPTER X: PRINCESS MAMSOUD OFFERS MARRIAGE

Much to our surprise, and joy, a relief force of Legion infantry took over the post at Mesme, and we were sent off to the town of Homs, for a rest. We were glad to see the back of Mesme. It had been the most stupid blunder to send a depleted squadron of 130 men there, and expect them to hold the post against the hundreds of Druses known to be massed in the area. I, for one, left the place with singularly ghastly memories. That decapitation was too close to me to be forgotten easily. And there was Scora, sent on to the madhouse at Damascus, never to see day light again. . . . He might have been a murderer in his own country. In the Legion he was a courageous fellow, a reliable chap, and a good friend. ... A rotten blunder from start to finish, and only to be explained by the mentality of the French authorities, who, of course, feel that legionnaires may be sent into any sort of hell-trap.

The rumor had got about, as rumors do, that we were to do a month's garrison duty in Homs, and then proceed to Rashaya. We were hearing quite a lot about Rashaya at that time. Apparently they specialized in particularly bloodthirsty battles up there, and there was a constant stream of troops moving into that cauldron.

In my somewhat eventful life there stands out with unusual clarity an episode that came my way during our stay at Homs. It was the sort of thing that can never be erased from memory, that happens to few men, that occurs to no man more than once in a lifetime. Even now, I think and dream, and wonder and cogitate on the possibilities of taking a big risk to go back there. . . . Though, God knows, I never want to see the Legion again as long as I live.

The memory is a happy one, perhaps the happiest of my life. It was at Homs that I was ridden down by a woman, a veiled woman. I was out exercising my mount, a rather difficult Arab pony that had been handed over to me to see what could be done about it. As a matter of fact, I did a great deal with it. We soon got to know each other, so that, finally, he was eating out of my hand in more than one sense.

I was giving him his head when another rider came along, careering across the open country at a terrific rate, and charged straight into me, almost knocking me out of the saddle. When I had gained my balance, I looked round, to find a veiled woman beside me. In a flash I drew my Mauser with every intention of firing, for there is no greater insult than this a native riding down a French soldier, whatever the sex or caste of the native; and the law to shoot without question applied even to legionnaires. But before I realized what was happening, she had spurred her mount into a gallop with the cry: "Oh, English!" and was racing across the plains as if pursued by ten thousand devils. She certainly was pursued by one devil, for I was simply crazy with rage. And all the time that laughing cry of hers, "Oh, English!" was singing in my mind. There was no doubt about it, she had spoken those words as one who knew English.

Which made the mad race all the more exciting, naturally! And she certainly gave me a healthy gallop. Long before I had caught up with her my rage had cooled and I was perspiring freely, so, too, was my Arab mount, the once-fractious and untamable brute! As a matter of fact, I don't suppose I should ever have caught up with her had she herself not given me the opportunity by drawing in her mount. She was that sort of rider.

I rode up to her until our saddles were almost touching. Then k leaned forward, trying to penetrate the thick, white veil, but she drew back and held up her hand. It was, I saw, encased in a neat leather riding glove.

"We have met before," I said, taking in the highly- polished top boots and the scarlet breeches.

"I did not know then that you were English."

"How did you find out?" I asked, marveling at her accent.

"It was very easy, Mister Legionnaire. When I ran you down, you said 'What the Hell!'"

"Did I? Yes, but . . . who are you?"

She laughed, a lovely laugh that caroled its way up and down the scales.

"Does it matter?"

"It matters a great deal. Do you understand? I should shoot you for riding down a French soldier."

"But you will not!"

"No?"

"No," impudently, "you will not shoot me."

"Why not?"

"I am your good friend. I have been searching for you!"

"The devil you have!"

"Yes. It is true. The devil I have!"

"But . . . why?"

"You have too many whys, Mister Legionnaire." "Where did you learn to speak English?" "At the American Mission in Beirut."

"Oh. So you were educated there?'*

"Yes. Three-four years. And my brothers, they too, educated there."

"Were those your brothers who were with you at the village of Mesme?"

"Two of my brothers. One more is in America."

"Really! What for?"

"To learn American business."

"What are you—you and your brothers?"

"Now, Mister Legionnaire, you have too many whats!" "Arabian?"

I persisted.

"No! No! No! Not Arabian!"

"Turkomans? Greek? Bedouin?"

"No! No! No!"

"Spanish? French? Portuguese?"

"No! No! No!"

“Raise your veil and then I shall know."

"No, no, no, no, no!"

"You have too many nos, Mam’selle Mystery."

And she laughed again . . . soft, lovely, carolling. Could she be old and ugly behind that veil, and yet possess such a soft, sweet voice?

"You will not raise your veil—then take off your glove."

She displayed a small, beautifully molded hand, the beringed fingers manicured and pink-tipped. She could not be old to possess such a soft, almost child-like hand. I counted ten rings, exquisitely chased and jeweled.

We looked at each other, at least, I stared at the thick, white veil, and wondered. I felt a sudden mad impulse to snatch at the white material and tear it away. But I thought better of it. I am glad now that I did not give way to that impulse. It would have ruined everything. By this time I was too deeply enthralled, too fascinated by the mystery of her. Who was she? What was she? Should

I tear away that veil merely to find an ordinary European woman . . . masquerading? Or some appallingly disfigured girl?

"I know what you are thinking—that I am old and very ugly.”

"No. Not with hands like this."

"Ah! But the English are not cold?”

"Tell me who you are."

"If I told you, would you come to my house—eat and drink with me? Or are you afraid?"

"I am not afraid," I said rashly, "of course I should come."

"Even when I tell you who my people are?"

"Who *are* your people?"

"Promise, Englishman, promise!"

"I promise."

"An Englishman does not break his word?"

"No."

"Never?"

"Never! But you are very exasperating. Tell me!"

There was a long, long silence. I waited. I was determined not to ask again. I just stared at that thick, white veil, wondering what was happening behind it. An irritating situation indeed!

Then she said it!

"My people are the Druses."

"Good God! But don't you understand? We are enemies.

The French are at war with your people!"

"But you are English. You have no quarrel with the Druses."

"I have no nationality. I am a legionnaire. Don't you understand? If I were found with you, I should be shot, shot as a traitor!"

"And who shall find, Mister Legionnaire, who shall find?" "'The Partisans . . . if we are not very careful. Those devils are everywhere."

"Partisans! Tcha! Dogs! Sons of Allah who have become Christians! Tcha!"

"And you have been educated by Christians, too!"

"Yes, it is true. I like the Christians who are American and who are English. The Druses have no quarrel with them. The Druses are happy for the education of the Christians; but they are still Druses, are they not? They do not make the pretense. They say, I like your Christian education but I am Druse. I am always Druse!"

"I understand. But I am a legionnaire."

"No, no! A little time you are legionnaire. But all the time, in your heart, you are English, yes?"

I nodded. What could I say to such logical reasoning on the part of this educated maid from the Jebel Druses mountains?

"What is your name?"

"Princess Mamsoud."

"No!" I gasped, forgetting myself in sheer astonishment, for I knew that no Druse would lie about his race, his caste or name.

"What is it you mean—*no!*"

"I was surprised, that is all."

"And you believe?"

"But, of course!"

"Now tell me your name?"

"John!"

"Ah!" she laughed. "That is it—John Bull, yes?" I

nodded, and thought it quite a good one.

"And you have promised to come to my house. You will come?"

"But I do not know your house. And there are your brothers. They would not welcome a legionnaire."

"They would not know. I show you . . . one-two mile now. You come to my house to-night, yes? Are you afraid?"

"I am not afraid. And you will raise your veil . . . for me?"

"I promise, John Bull. You shall see. Now come! I show you where is my house, and none shall see you."

I have done a few crazy things, quite apart from joining the French Foreign Legion, but following that girl I had made up my mind that she *was* a girl may be classed as about the craziest. It should be remembered, however, that I was very familiar with all manner of Eastern types of houses, having been invited into some, and having thrust myself into scores of others.

This, though, was in vastly different case. I might well end the escapade with a knife in my back. Not that that worried me a great deal. A legionnaire lives so perilously near to the verge, has to risk so many things, murder, sudden death, mutilation, has to endure so many hardships, so much callousness and cruelty, that he becomes something more than hardened. He becomes a fatalist. I am a fatalist to this day. It is, once thoroughly ingrained, the most comforting philosophy I know.

No. It was. not the knife in the back and no chance to retaliate. It was the fact that I might be making a fool of myself, might be allowing a woman to lead me into a trap. I had been led before ... by women. One led me into the Legion ... or, at all events, pushed me there. Still, I was, and am, incurably romantic, not to say human. What would you? A couple of years or so in the Legion at that time, and nothing to look at save cheap and dirty Europeans and cheaper and dirtier natives of Africa and Syria. This is not a justification of an act. It is not even an excuse. It is an explanation.

Princess Nashalla Mamsoud was the daughter of a Syrian Prince, and a direct descendant of Sultan el-Atrash, of the Jebel Druses. Hers was a noble family, as noble that is, as any of the race could be. Certainly she was wealthy. I came to know, later, something of the extent of her wealth; and even I, knowing something of what the Easterner can do in the matter of wealthy display of ornaments, jewels and the like, had occasion to gasp.

Well, not to put too fine a point on it, I visited Nashalla towards midnight of that day. Everything had been arranged. I wore a burnous over my uniform and sandals on my feet, these being handed to me at the house of one of Nashalla's servants, when I reached the native quarter. We then wended our way through a labyrinth of dirty alleys and passages until we came to the courtyard of the Princess's private apartment. I was left crouching in an archway, feeling all sorts of a fool, while the servant, a trusted negro, went forward to investigate.

Apparently it was not going to be too easy. He must have been reconnoitering for fifteen to twenty minutes, though it looked like as many hours to me, before he returned to say that all was well. The quarter was pitch black, for the overhanging buildings obscured what little light came from the night sky. In that mysterious dimness I had to hang on to the belt of my companion, for it was impossible to keep him in sight otherwise.

We came to a heavily-studded door in what appeared to be a blank wall. My guide whispered a few words to a couple of armed natives who passed us through. Keeping close to the wall, we shuffled round the courtyard, finally coming to another door, which led into the building proper. Here again there was a whispered consultation . . . and then I found myself inside a room with my guide, my hand closed over my Mauser. I stood with my back to the door, watching him as he proceeded to light the candles.

As the room gradually brightened, I became aware of its furnishings. The floor was covered in a thick pile of rugs that deadened the sounds of movement. A low settee smothered in gay cushions stood near one wall, and beside it was an ebony table holding bowls of fruit and flowers and sweets. But it was the walls of the room that held my attention. I have seen some wonderful things in the East, but never interior walls quite like those. They were covered with colored tiles cunningly set to form pictures, and the dancing light and shadow from the candles seemed to give those pictures weird life and movement. On one wall a whole army of Druses appeared to be moving across a real desert, while the candle lights picked out the colorings of their clothes and arms and the trappings of the horses. Another wall was blatantly sexual, holding the sort of picture that could never be tolerated anywhere except in the East. Still another was set with a picture of one of the famous fountains I had seen in Damascus, around which were grouped a number of children. This wall was partly obscured by heavy curtains.

The negro indicated the settee, and I sat down deeply into a pile of cushions. Nestling under one of the cushions was a huge box of cigarettes, and since I had already made a fool of myself, I took another risk and started to smoke one. In just those surroundings it tasted beautifully cool and sweet. The negro muttered something unintelligible to me, and disappeared through the curtains. Presently he returned, passed through the room and out at the door by which we had entered.

As soon as he had gone the curtains parted again. Princess Nashalla stood there, looking to my excited eyes like some vision from an unreal world. She still wore the thick white veil, but had discarded the top boots and the scarlet breeches. Now she was dressed in a long robe of white silk that made her seem exceptionally tall and graceful. She came soundlessly towards the settee and sat down beside me. The white silks fell apart, revealing pink heelless slippers and satin legs of the same color. Her ringed hands closed over mine.

“You have courage, John Bull.”

Her tones were low and soft and intimately sweet.

“You thought I should not come?”

She shook her head. “I knew that you would come.”

A pause. She drew my hands forward until they were clasped behind her, and I was appreciably nearer the soft silks. Even now I can smell that delicious perfume. It was the most delightful thing that has ever assailed my nostrils. I do not know what it was, and I don’t suppose I ever shall. It was almost overpowering, coming up from the warmth of her, drugging the senses.

"Will you not take off your veil?"

She did so, with my face a few inches from her own. Heavens! It was the face of a child almost. Sixteen she had said, but maybe she had been trying to forget her girlhood. Her hair was dark, not black, with partially hidden flames of color. Her eyes were deep and dark, alluring, gay and sad by turns. She had the full red mouth of a child and the eyes of someone very old and very wise. She might well have passed for a daughter of Europe, rather than a native of the Near East, for her skin was too fair, too apricot colored. The lifting of that veil had come as a shock to me. I had seen just such features and coloring on the streets of London, nay, darker more olive-tinted features. And I fancy there was just the faintest glimmer of disappointment in that revealing of a face, since it was not in the least as I had visualized it during the whole of that blessed day.

Nevertheless, the girl was a lovely allurement. The generous red lips of the childish mouth were ripe for kisses. I cannot say how long we sat there, gazing at each other, I taking in the wonder and beauty of this creature who had come to me, a legionnaire, out of the plains. I could have sat there for timeless periods, staring upon that exquisite picture. Slowly the full lips parted, showing beautifully even, ivory white teeth. The smile filled her dark eyes, lighted her features.

Yes, I thought, she knows I am deeply impressed, knows why I am staring, speechless. I admit the beauty, but it is beauty not of the East. It is the beauty of Southern Europe.

An amazing people, those Druses. And, after all, Southern Europe is not a far cry from Syria.

"Now," she whispered, "are you happy?"

"Happier than I have ever been before ... or ever shall be again."

She pouted. "But why not again . . . and again? Why not always happy . . . now?"

My God! A legionnaire always happy! I was suddenly overwhelmed with a sense of bitter depression. I realized how utterly impossible was this strange affair. In a few hours I should be back in the garrison, back among the foul-mouthed, butchering scum who were my comrades in the Legion of the Damned, a Hell Hound of France who was paid three-ha'pence a day to slaughter this girl's kinsmen. Could this, by any stretch of imagination, be called a happy state of affairs?

Evidently my fatalism was not so deeply ingrained as hers, for I could still be troubled by the thoughts of to-morrow. That tomorrow butted in, shattering the ecstasy of those moments, caused a cold fury to sweep over me. For I knew, even in that first encounter, that it was to be no ordinary sex adventure. This was different. The spice of danger, the risks, the hazards—they were nothing at all then.

Always happy! In a lust of blood and drink and cheap women!

She was watching my face now, a frown of perplexity rippling the broad level brows.

"Oh, why are you not happy with me, John Bull?" I

laughed suddenly.

"I was thinking how impossible it all • . . . this meeting with you . . . here. Soon I must return to the Legion. In a few days I may go with the Legion ... to some place . . . perhaps Rashaya . . . and there I shall kill your countrymen!"

"Oh, John Bull! John Bull! Do not speak of these things! Am I not beautiful, that you should talk of killing? Are you not happy, to be thus . . . with me ... in my home?"

"It is because I am so happy, here, with you ... I must go."

"No, no, no, no!"

And she clung to me, her soft mouth pressed against my face. . .

.

"I would speak much with you . . . about the Legion."

"Yes?"

"Do not go now. Wait! Wait! Do not go back to the Legion! Not to-night, nor to-morrow, nor ever again! Stay with me!"

"But that is impossible! I should be found and shot! "

"No, no, you would not be found. My brothers know! I have spoken of you, the Englishman, John Bull, to my brothers! "

"The devil you have!"

I jumped to my feet and made for the door. I drew my Mauser and turned to face her.

She tore the white silks from her shoulders, exposing the soft breasts.

"If you think I have betrayed you shoot!"

I shoved the pistol back again, feeling I had made a frightful exhibition of myself.

"But your brothers?" I queried, wiping the perspiration from my face.

"Are you afraid of my brothers, John Bull?" "No. I

am not afraid of any man—out in the open." She

drew me down to the couch.

"I have spoken with them. I have said you are English and my very good friend. They understand. Why do you not understand? "

"I do understand you . . . now. I am sorry. Of course, I trust you. But I do not know your brothers."

"Will you shake the hands of my brothers, if I bring them?"

"When? Now?"

She nodded, smiling.

"No," I said, "not this night."

God in heaven! How could I shake the hand of a Druse? The situation was becoming more and more impossible. I wanted this girl—like hell I did! And I knew it for an insane want. Try as I would, I could not adopt the wineshop manner with her. Nor could I associate her with the Druses, the enemies of the French, and therefore the enemies of a legionnaire. What was the good of calling me English, and John Bull, John Bull, John Bull all the time? I never knew whether to laugh or to weep every time she said that ridiculous name.

But, deep down in me was the feeling that this girl was not really of the Druses. The feeling had no sort of reasoning behind it, except that her face was not the face of a Druse woman. Did I not know the appearance of a Druse woman? How long was it since I had taken pot-shots at them as they hung, feet uppermost?

"Then you will go back to the Legion this night?"

"I must."

"Why? Would you not leave the Legion . . . escape?" "And fight for the Druses?" I sneered.

"No, no! No more fighting. Come with me to America!" "What!"

I shook her until she said it again, laughing and gasping.

"My brothers would help us. I have spoken with them. We could go with them to Palestine. You would wear the burnous, just as they. No one would stop the Jebel Druses, not even the Partisans. We could travel in Palestine just you, John Bull, and Princess Mamsoud. Then we should go to America. Did I not tell you that one of my brothers is in America ? " I stared at her in amazement.

"But this is madness!'* It

was her turn to stare.

"Madness? Is it madness to be happy? Are you not happy with me?''

"I am too happy with you. Do you not understand? We should need much money to travel to Europe, to sail in ships.''

She laughed. "But I have much money, too much, too much!"

The situation was becoming more and more impossible. I felt a growing sense of unreality, as if it were some crazy play in which I was taking the major role.

She left the room, asking me to wait. In a few moments she was back again, carrying a map. It was an American map, evidently one she had had at the Mission in Beirut, and she insisted on showing me the route from Beirut to New York, the way her brother had gone. Obviously she had made up her mind to go to America. She had all the details worked out. She knew what it would cost. Apparently her brother kept her well-informed about the conditions over there, and he had certainly fired her imagination.

“You will come with me, John Bull?” she begged, smiling up into my face.

“I have no money.”

“And I have much money!”

She took my hand and made as if to draw me towards the curtains.

“Come! Let me show that I have much money, much jewelry!”

"No,” I said. "I will come again to-morrow night. You shall show me then. And we shall talk about this.”

“You *will come?* And you will marry me . . . here . . . in my house to-morrow night?”

"I will come to-morrow night,” I said.

And she had to be satisfied with that.

CHAPTER XI: FLOGGING—AND LOVE

I did not visit Nashalla the following night, nor the night after, nor the one after that. I was a sick man. I was indeed a sorry sight for so lovely a creature as Nashalla to contemplate. I had endured a particularly damnable dose of flogging. Instead of seeking out that vision of loveliness, I lay writhing with a lacerated back, the flesh of which was almost in ribbons. McCann, splendid fellow, was as solicitous as any trained nurse. He did all he could to clean the salt from the cuts, to bring some relief against the excruciating pain.

Instead of visiting that loveliest and sweetest of women, for she had become all that to me, I lay writhing and cursing. Instead of those anticipatory pleasures the planning and scheming for escape and flight to America I cursed the French until I could curse no more. I could be as foul-mouthed as any other legionnaire. And I was. How I blasphemed the French and all they stood for! McCann has said, in all seriousness, that it did him good to sit and listen to it.

And I can still do it. I hope I have made that clear. My hatred of the French, born in those fiendish and ghastly conditions, has not abated one iota. I hate and loathe and despise them as much today, in this centenary year of their hellish Legion, as ever I did while a Hell Hound in their pay. I am not suggesting that my puny hate is likely to cause them any sleepless nights. A nation that can perpetuate and encourage such diabolical systems as the French Foreign Legion and

Devil's Island is hardly capable of appreciating deep human emotions, of which hate is an essential one.

I can imagine their amusement.

Nevertheless, hate has its uses. It is a spur to the emotions, to the mind, to the memory. It brings a clarity to the truth that will not be gainsaid. I had been flogged before. I was flogged again and again, before I finally escaped. It was not a particular flogging that bred hate. It was an accumulation of floggings and cruelties and barbarities.

I single out this flogging because it was undeserved, even according to Legion laws, which is saying a lot. Moreover, it was a deliberate playing into the hands of one who had long sought to spend his muscular strength on my back. To wit, one Sergeant Lopez, of French-Polish origin. His opportunity came during the temporary absence of Captain Dupont.

This N.C.O., as much a victim of the system as I, no doubt, had waited by my bunk for nearly five hours, and when I arrived, fresh from the loveliness of Princess Mamsoud, he was overjoyed at having to throw me into prison for being absent without leave. He was even more overjoyed when I was awarded a flogging, and he the administrator. He could never understand why McCann and I were favored by Captain Dupont on sundry occasions, by that same Skipper who was callous and brutal to all men. Had he been through as many "tough breaks" as the Skipper, Mac and I, he might have understood that even so brutal a man as Captain Dupont could appreciate tried fighters.

And so, this sergeant of French-Polish extraction, who had the figure and the strength and the brains of a gorilla, made the most of his opportunity. In flogging me until I lost consciousness he was quite within his rights, and he certainly exercised those rights. He did not fail to express his sympathy for me as he wielded his whip so expertly. He thought it a shame that any girl of the wineshops should have kept me until within a few minutes of morning parade.

He died, later, in very mysterious circumstances. But those circumstances are part of a later section of this chronicle.

He had his day. Every dog has his day. Even dirty dogs.

I do not hate him, poor devil! How hate a dead man? He was but part of an organization that is rotten from top to bottom. The system bred and fostered such brutes and continues to do so.

Hence, four days passed before I again ventured into the native quarter as far as the house of Nashalla's trusted negro servant. No one knew where I had been on the previous occasion, not even McCann. It was assumed that I had been indulging too heavily at one of the wineshops, and I let it go at that. McCann, of course, was mighty curious. It was the first time during our companionship in the Legion that I had gone off alone and in such a secretive fashion.

"Waal, Barrington, I guess you've got hold of some nice cup o' tea that doesn't offer its wares in the usual dives. But you're a secretive cuss, and I think you ought to let a pal in on a good thing." "McCann, you're wrong. She's not a nice cup o' tea at all. She's straight."

"Yeah? An' all for a legionnaire? Shoor! I guess I've heard of the brand before."

"You know what you felt about that girl at the *Cafe Noir?"*

"Don't mention that bitch to me!" he spluttered. Evidently the incident still rankled.

"Well, Mac, I may be wrong, but that's how I feel about it."

"O.K. with me, boy. But, for the love of Mike, watch that son of a bitch, Lopez. He's still thirsting for blood." And then, I suppose, I spluttered.

"Anyway, now the Skipper's back, I don't think he's too happy, Mac."

"I guess he'd like to change his squadron . . . before it's too late," growled McCann mysteriously.

It was sheer madness, of course, but at that time I thought it worth risking even another flogging. After all, one does not pay court to a Princess every day.

The amazing thing about this second visit was that, though I had not been able to warn Nashalla of my coming, I found the negro waiting for me with burnous and sandals, and when I had put them on we strode off without a word being exchanged. She had guessed, then, that I should appear tonight

"I knew you would come this night," she said, when we were seated together in that picturesque abode.

"How did you know?"

"I knew that you were beaten with the whip, and I knew that you would not tell of your visit here." I sat up and stared at this sorceress.

"How did you learn these things?"

"The Druses know many things about the Legion. They talk of these things in the bazaars, of how the French beat their legionnaires."

Her soft arms crept round my neck.

"Will you not kill the Sergeant Lopez," she whispered.

"No," I said. "How can I do that? It was the officer's command that I should be flogged."

"Will you not kill the officer?"

"Listen, Nashalla, we must not talk of this thing. It is finished. You must forget it."

I had a strong suspicion that she wanted the officer's name so that she might offer his killing to one of her kinsmen. And that might well have started trouble in Homs. I tried to explain this to her, knowing that killing was a small matter in the philosophy of the Druses.

She shook her head, a strange smile on her face.

"You English have much courage. But you are sometimes strange."

Then she began to talk of marriage and of our flight to America. She seemed to have no doubt now that I should fall in with her wishes, though she did not press the matter, nor did she expect me to marry her "this night." Without saying so in so many words, she managed to convey her appreciation of my condition. She guessed that I was still sick and sore. And I certainly was. One cannot get over that sort of punishment in a day or two. All the same, knowing something of her race, I could not help marveling at her sympathetic understanding.

"But, come! Let me show you!"

She took my hand and led me from the room, along a dimly- lit passage, and down a steep flight of steps. When we reached the bottom we were met by an old crone of a woman, who came shuffling towards us, a huge bunch of keys jangling by her side, a lamp in her hand. The old dame was unveiled, her face a mass of wrinkles, yellow and dry as parchment, the eyes deeply sunken. She looked like a hundred years or more, and she was bent almost double. A weird scene indeed! The one so young and lovely and desirable, the other so old and shrunken and lifeless!

With much fumbling and muttering, the old woman selected a key from the huge bunch fastened to her girdle, opened a great heavy door, pushed us inside, and locked it again.

“Well,” I laughed, “I hope that old dame will let us out again. She did not seem very pleased to have us down here! ”

“Look, John Bull that is why she is a so suspicious old woman!”

I was looking. I was staring in astonishment. My eyes were popping out of my head at the sight before me. The floor of that vault was literally covered with sacks and chests, and some of the sacks were open, spilling gems. We were in a sort of jewel storehouse. I have never seen so many precious stones in my life. I dug my hand into a sack and brought up as many as it would hold of blood-red rubies. Heavens! How I stared! Nashalla began to laugh, seeing my open-mouthed admiration and astonishment. "You cannot believe your eyes, John Bull?"

"But, of course," I said coolly, letting the delicious things trickle through my fingers.

I went to another sack and did the same hundreds and hundreds of pearls. Then another of emeralds, green as the shimmering shallows, agate, black as night, amber beads, cat’s-eye stones, and a score of other varieties that I could not even name.

"Well," I said, trying my hardest to appear nonchalant, "you have a nice collection of stones here."

But Nashalla did not appreciate sarcasm. She took some keys from the folds of her robe and unlocked chest after chest. Loot, of course; but, my heavens, what loot! The collection of many years. I knew a bit about loot. I had several hundred pounds' worth of my own. I knew a jeweled saber or dagger when I saw one. I had seen so many in various parts of the East that I had become something of an expert at valuation. The ambition of all legionnaires of experience is to collect as much loot as possible and desert; but the chances of gathering loot are far greater than the opportunities for desertion.

Anyway, these Druses had us all guessing when it came to looting. Where we purloined hundreds they got away with hundreds of thousands. If one-tenth of that treasure were to be dumped in Hatton Garden the London market would collapse immediately! And that was not all. Much more loot had at some time or other been changed into currency. In one chest were bundles of British, and French notes and American bills. It was then that I began to appreciate why Nashalla's brother had gone to America to learn business. But what a risk, what an insane risk, to show all that loot to a legionnaire! I turned to Nashalla and said as much.

And she laughed. Lord! How she laughed! The white teeth flashed. Her eyes danced with merriment. Apparently it was the greatest joke she had heard in years!

She was irresistible. The soft silks of her shook and shook in my arms. The dancing eyes, the bewitching mouth came close, while that carolling laugh rang its way up and down the scales. There was not a stone there in all those thousands that could scintillate as she did in those delicious moments. Just to hold the laughing creature, shimmering with love-light and brimming with vitality . . . was ecstasy!

Presently she subsided, save for an occasional ripple, the soft sobs of dying laughter, and lay calm, her eyes shining . . .

We passed from the vault into another, and in that other was a score or so of Druses, all of them armed. They stood aside hurriedly as we passed through, and there Nashalla showed me a very ingeniously arranged cofferdam, a sort of heavily-built stay that kept back the water of an underground river. In a few minutes the vaulted parts of the building could be flooded to the river level. No wonder she laughed.

We went back to the room of tiled pictures.

"You believe now that I have much money, yes?" And

the laughter was still in her eyes.

I picked up the box of cigarettes and offered it to her. She took one, applied a light, pushed it into my mouth, then took another. And never was a cigarette so sweet.

While we smoked, seated close on that couch, she talked about our marriage, which was to be in the Druses ' mode, of our coming travels, the wonderful, wonderful days that lay ahead for us. And because I am a person of some imagination, she must fill my mind with roseate visions and pictures of a glorious future until my head swam.

In that hour, abounding with passions and promises, I felt I could desert ten thousand French Foreign Legion, overcome any and every kind of obstacle, travel to the ends of the earth in the company of this altogether lovely and desirable creature.

She jumped up from the couch suddenly, kicked off her slippers, and began to pirouette in the center of the room, laughing softly, tantalizingly, provocatively. Her dance was not the dance of those I had seen in the Pit and other wineshops. If it had been, I should have hated it. She danced for me, and it was utterly graceful, dreamlike and beautiful. She was a white flame, darting fiery essence about me, so that I burned in answer to her fire . . .

That sinuously graceful, exquisitely slender, figure tore at my sense of reality, as no one has ever done, before or since.

I can see her now!

If only one might turn back the hands of time!

How different life might have been!

CHAPTER XII: BATTLE OF RASHAYA

"For the love of Mike, Barrington, what's bitten you? Ever since we left Homs you've been going around like a bear with a sore head! You're the perfect picture of death warmed up! You look about as comfortable as a duck in a desert! If you was not thoroughly smitten by some nice cup o' tea in Homs—waal, I'm Queen Victoria!"

"Oh, go to hell!"

"Waal, boy, I reckon I shouldn't have far to travel!"

That conversation took place at a time when I couldn't stand the sight of anything in French uniform, not even a legionnaire, not even my pal, McCann. We were *en route* for Rashaya, proceeding through mountain passes that ought to have aroused my admiration and delight, for there was a rugged beauty and grandeur about the country that was good to behold.

I was fed, utterly fed! Soon we should be knee deep in murder again, and I did not care how soon. My wonderful romance had gone west, flopped thoroughly just at the time when I was on the point of getting away from the Legion forever. I might never see the lovely Nashalla again! Of course, my mind was full of her, wondering what she would think of my sudden disappearance.

"Oh, John Bull! Then you were afraid?"

Would that be her plaint? I wondered. A wave of sickening depression descended upon me. I could not shake it off. Had I been a jilted youngster I could not have felt more sick! It was the vilest piece of luck imaginable. I had visited Nashalla again, had spent another joyous night in her picturesque abode. Our plans were complete, arrangements had been made for my final appearance on the following night, our marriage according to the rites of the Druses, and our escape to the Palestine Border. I returned to barracks that night, or rather early in the morning it was about three o'clock with my head full of roseate schemes for our future. For me the universe had suddenly burst into song. And then the bombshell! I walked into barracks at three o'clock in the morning to find the whole of the Fifth Squadron tumbling out of their bunks. Even then, I should have made the attempt to get away if there had been the slightest possible opportunity I should have turned there and then and doubled back to Princess Mamsoud.

But there was not a ghost of a chance. My name was being shrieked around the place. The Skipper was bawling himself hoarse. As soon as I showed myself Sergeant Lopez pounced upon me, and I was kicked all the way to the Skipper's quarters.

"Barrington, *nom du chien!* Cursed pig! Why do you not come when I command? "

I spent the next hectic half-hour getting the Skipper ready for our march up to Rashaya. As it was, McCann had to draw and saddle a mount for me. But all the time, as I rushed here, there, and everywhere, I kept my eyes open for the smallest chance of escape. What was the use? I hadn't a dog's chance. The whole camp was roused. A full column was to march out immediately, with orders to advance on Rashaya and keep a look-out for a band of rebels that had cut up a detachment of a hundred men and twenty- six officers.

We never found those rebels, but we came upon the mutilated bodies of the officers and men of that detachment. They were laid out, over a hundred of them, in the most bestial fashion, for our special benefit. The Fifth Squadron, heading the column, found them in a narrow gully in the mountain passes, the most filthy sight imaginable, every vestige of clothing and equipment gone, their naked, sun-scorched bodies hacked and slashed, some of them grotesquely veiled, and in such a manner as to leave no doubt regarding the brand of females who had assisted in the fiendish pastime.

We spent two fruitless days in the neighboring hills and gullies. Save for that frightful mutilation, there was not a trace of Druse or Bedouin anywhere. Practically the whole of that time McCann was continually raving about the marvelous mountain scenery. It reminded him of the Rockies in Canada, and he grew quite sentimental at times until I told him to shut up. I suppose it was beautiful, but I was not in a frame of mind to appreciate it. I had been robbed of my romance and my freedom by the rottenest fluke!

Should I ever see Nashalla again? And if I did what then? What would she think of me? Perhaps she would know of what had happened that night? She seemed to be pretty well informed of all that passed in the garrison at Homs. Perhaps she had expected me to double back that same night? Would she realize that I had not had the ghost of a chance? For three days I was nearly frantic with the queries that spun round and round my tortured mind.

So that I was not at all responsive to Mac's laudatory comments about the mountains that reminded him of the Canadian Rockies.

“For God's sake, Barrington, pull yourself together! There's going to be a lovely war as soon as we get to this goldamed place, Rashaya. Besides, I guess we'll find a nice cup o' tea somewhere around."

That night the column camped in the mountains. It was a wonderfully strategic position, high up, and overlooking vast tracks of country. Man and beast were worn out after the long day. I, for one, remembered nothing until dawn, when I was suddenly awakened to find the Squadron in a turmoil. N.C.O.'s were racing round, booting the men into wakefulness. Four of our comrades had been brutally murdered in their sleep, and in circumstances that were very suspicious indeed. The four were practically in the middle of the camp. It could hardly be the work of bandits, stray Arab looters or Bedouins. For such outsiders to have done the deed would mean that they had escaped a number of sentries, ignored a lot of sleeping men, and picked on these four for no reason at all. Why should they have taken such risks, when there were easier marks on the outer confines of the encampment?

Obviously this was not the work of natives at all. The murders had been committed by someone inside the camp. Four of our comrades had been brutally and mysteriously murdered by one of our own men.

Captain Dupont went around like a raving lunatic. He swore he would get to the bottom of the mystery, if it meant sending the lot of us to the penal battalion. Four of our comrades had been stabbed through the heart. Why? Where was the maniac? Who was he? The Squadron was paraded. Every man and every officer was turned out. The Skipper made exhaustive inquiries. Everybody was completely mystified. There did not seem to be a clue of any sort. Two Germans, one Dane and one Russian, all good soldiers of the Legion, had been murdered in cold blood for no apparent reason.

The whole Squadron was nervous, uneasy. Some of the men thought it must be the work of bandits, since there was no reason why a comrade should do this thing. But why should bandits pick on four men sleeping in the middle of a camp, men who were entirely surrounded by other legionnaires and by sentries? On the other hand, why should one of our own men have done this thing? One strange fact was revealed. The four men were chums. They had been in the habit of going about together whenever the Squadron was resting. Then murmurs began to float round that the men had not been murdered, that this was a suicide pact. If it were, it must have been a rather gruesome one, carried out with great speed and dispatch, and very quietly which hardly seemed feasible at all.

The Skipper was far from being satisfied. We buried the men and started on the march that day in a queer state of mind. The officers were conscious of the restiveness of the Squadron. There was mutiny in the air. If four men could be murdered like that in cold blood, in the middle of a camp, at dead of night was any man safe? We were subjected to even more severe and brutal discipline, and there were several floggings, before we finally got on the march that morning.

As we proceeded we went higher and higher into the mountains, and by late afternoon we came upon Rashaya, high up, towering, as if it were set on top of the world. A well-made road led up to the town. There were signs everywhere of orderliness and good planning. Without doubt the French had taken great pains with Rashaya.

It is about the most modem town there is in the Near East. The Citadel stands on an eminence overlooking the whole place, and is approached by a very steep and narrow road.

The huge timbered gates of the fortress were thrown open on our arrival. We were greatly impressed, for the place looked well fortified and practically impregnable. Companies of Spahis Moroccan and Senegalese were lined up to receive us. It all appeared very promising, and we thought our stay should prove decidedly pleasant. One company of Senegalese was left behind to share the duties of garrison with us, and the remainder of the troops marched out *en route* for Damascus to rest.

In the following two or three days I was able to review my own problem a little more calmly, for our garrison duties at the Citadel were comparatively light, and soon we were moving quite freely about the area. I began to hope, to dream and scheme. I must, I felt, prepare for the day when we should be near Homs again. That day was bound to come, since we should eventually return to Damascus. I determined that if ever we came within reasonable distance of Homs I should make a bolt for it. It did not occur to me then to wonder whether Nashalla would still be there since she had done a great deal of travelling and if she were not, what would be my next move.

One of the things that marred the peace of those first few days in Rashaya was the persistent and brutal animosity of Sergeant Lopez towards McCann and myself. It became so insistent at times that Mac and I discussed what we should do about it. It was even worse against McCann than myself, as, being the Commandant's servant, I did not get in his way so much as McCann seemed to do. Just when we thought we were all set for an hour or two at the particular dive we had found, McCann would be doing extra drills, or guard, or we'd both be burying dead anything that was dirty enough.

The climax came through the Skipper himself. I thought at first that Captain Dupont was trying to associate himself with Lopez in leading us a dog's life. I realize now that was not so. His instinct was quite the reverse.

Rashaya at that time was beautifully quiet, so quiet that we had time to investigate the murder of a native interpreter in the town. The Skipper sent McCann and me with Sergeant Lopez of all people to look into it. The Sergeant rode ahead and we tailed him, telling him what we thought about him but not loud enough for him to hear!

"Barrington," said McCann, "I'm going to get quits with Lopetz before I'm much older!"

"Don't try anything on to-day. . . . Wait a bit."

"Wait nothin', you poor sap! It's now or never. I can't stand the little runt any longer!"

"Be careful, Mac. He's the finest knife-thrower I've ever seen."

"Yeah? And we are the finest shots in the Legion. What say?"

"Count me out. It's too risky, with only the three of us out on a stunt like this."

We found the interpreter's house, and the Sergeant made some inquiries about the killing. He learned precisely nothing, but he decided to stay at the house for a little while. Apparently the attractions of the departed interpreter's daughter appealed to him, and he ordered McCann and me to take a look round the neighborhood and make further inquiries.

I suggested to the Sergeant that he should exercise great care, since there had already been one killing in the house; and for my pains I got a dirty cut across the face with his whip, with the injunction to mind my own business. The blow nearly knocked me from my horse. It was so unexpected that I was dazed for several minutes. When I came to, McCann was bathing my face, and the Sergeant had disappeared.

"Waal, can you beat it, boy? That swob in there, with his nice cup o' tea, says we must report to him in an hour. D'ja still love him? Isn't he a nice man to go riding with on a pleasant Sunday afternoon? Wait! You son of a gun, Barrington, he'll be lucky if he sees to-morrow's daylight!"

I agreed. In those moments I was ready to go into the house after him, and put paid to his account there and then. Instead, we wandered round this quiet backwater of the town, found a number of Christian Arabs here and there, and made a few hopeless inquiries. The result was nil, of course. One Arab offered us food and wine in his house, but I shook my head, telling him it was forbidden. Mac, however, insisted on accepting the hospitality. So it was arranged that I should keep a look-out for the Sergeant while McCann indulged. I was not at all in the humor for either women or wine.

I rode round for a while, rather enjoying the peace of that lazy afternoon, with only my own thoughts for company. I need not say of what my thoughts were composed. Nashalla was still very much in the forefront of my mind. I must have wandered rather aimlessly about, for I lost all sense of direction, and suddenly realized that I was on the outskirts of the town. The next half hour was spent in trying to find my way back, and a hectic time it was!

After a great deal of searching, I came upon the house of the interpreter, and was just about to make off in the direction that Mac and I had taken, when I heard a terrible scream coming from within. Dismounting, I dashed into the house, and almost stumbled over Sergeant Lopez.

God! What a sight he was! He was spread-eagled on the floor, a knife buried to the hilt in his neck. He was still breathing as I removed the blade, but in another moment he gave a horrible gasp and was dead. I ran through the place. There was not a soul in sight. In the courtyard nearby I found several Arabs dozing in the shade. One of them had the bridle of the Sergeant's horse fastened round his arm.

They knew nothing. They had heard nothing. Neither were they in the least moved by the sight of the bloodstained face and neck. Then McCann came on the scene.

"Suffering Mike! How did this happen, Barrington?"

I stared at my chum for a long minute, wondering if Mac were just a damned good actor.

"God knows. I was riding past the place when I heard an awful yell. So I got down to have a look. This is how I found him."

"Where's the bird?"

"I've not seen her. The place is deserted." I

swung round suddenly.

"Have you seen her, McCann?"

"Are you crazy? Where in hell should I see her?"

"All right. Let's go."

We covered up the body of Lopez, strapped it to his mount, and rode back to the Citadel, where we reported the incident to Captain Dupont. He listened very attentively to our story, fired a number of questions, then stared hard and long at us both. After this uncomfortable scrutiny, during which a peculiar gleam might have been detected in his black eyes, he waved a hand of dismissal. And that was the end of Sergeant Lopez. True, a patrol was sent to make further investigations, but nothing came of it.

On the fourth day runners reported that the Druses were gathering in the hills for an attack on Rashaya. The Commandant sent out a party of men under Lieutenant Duprez to reconnoiter. The peaceful town, inhabited largely by Christian Arabs, changed overnight. Women, girls and young children were begging to be taken into the Citadel. Many of them left their menfolk and followed the troops into the fortress, and finally Captain Dupont decided that all women and children should be accommodated. It was impossible to take the men as well. There simply wasn't room for them. Here, indeed, were new conditions of war for the legionnaires. The women and children had to be fed and housed in some fashion, but obviously arrangements for their welfare were anything but satisfactory. Most of them were simply dumped into the barrack rooms and left to fend for themselves. When the boys were not playing with the children or running after the women, they were being flogged and kicked into a semblance of order and discipline.

It was a chaotic situation to begin with, and through it all I still have a very vivid picture of McCann, with a child on each knee, singing the "Star-spangled Banner/' much to the youngsters' delight, for, just as women are the same the world over, so are the children.

Suddenly there came the echo of shots. Far away on the plains below we saw the reconnoitering party trying desperately to get back to the Citadel. As they came nearer we, who were watching, saw five or six of them topple from the saddle, riderless horses began to stampede frantically up and down the plains, until brought to their knees by further shots. What was left of the party came galloping up the road towards our gates at a terrifying speed. Quickly they were hustled inside, and the huge gates closed and barricaded.

On the heels of that party came the massed ranks of the Druses, advancing from all sides. Immediately below us, in the town itself, the Christian Arabs were putting up a desperate fight to hold their homes, and when the Druses came within range, we added our quota from the fortress. Soon our machine guns were spattering and chattering over the heads of the Arabs, by now terribly thinned down and driven almost to the walls of the Citadel. Nevertheless we were holding the massed hordes of yelling beasts.

The noise and din of that battle was the most heartrending I have ever heard. Mingling with the wild cries of the Druses, "Yelia! Jo Allah! Yelia!" were the screams of women, the pitiful cries of terrified children, the cursing and moaning of the wounded. It was bedlam let loose with a vengeance! All day and all night it continued, a terrible wave of sound, falling for a space, only to rise again to a horrifying crescendo.

The battle was strenuous enough, in all conscience. But it was not the fighting that worried us. It was the incessant screaming of the women and the wailing of children, the infernal din in our midst, harassing the men as nothing else could have done. During the second day of battle the wailing was making the men nervy. What remained of the Christian Arabs we had by this time drawn into the fort, for our ranks were thinning, and the Senegalese, hopeless conscripts at best, had lost a big number of men.

Hundreds of Druses had been slain along the road to the Citadel. At last the enemy retired, taking a great deal of loot from the town and a number of Arabian prisoners. They fell out of range to plot fresh mischief. That night they spent in wild singing and dancing, and a weird wailing came up to us from the plains.

"Those sons of bitches are just the same as the old Indian warriors in some ways," observed McCann, "with their fanatical singing and dancing. I guess they are working up their spirits and beseeching the aid of old Allah for a renewed attack."

During this lull a number of us went over to the women and children, who were all sheltered in the underground barrack rooms of the fort. What a sight they were! The place stank of their presence. Many of the youngsters, poor little devils, had been sadly neglected during the past three days by their wailing and helpless mothers. We did what we could. Rations were handed round. Captain Dupont gave out a double ration of wine to the men which was perhaps a mistake with so many women about. Nobody knew what might happen before dawn, and apparently nobody cared. We had to do something to drown that gnawing, ceaseless wailing and yelling that had been dinned into our ears for three days and nights.

So we became frightfully drunk and began to sing too! God! What a racket it all was! A score or two of national anthems were being mouthed at once, everybody trying to drown the efforts of his fellows. The "Marseillaise," the "Star-spangled Banner," "Wacht am Rhein " "Kong Christian," "King Karl, the Young

Hero," "Wien Neerlandsch Bloed," "Szozat," and "Magyar Hymnusz" an incredible babel of sound, a frightful screeching of pale patriotism and blood. The children, clinging to their mothers, with chubby fists full of beef and black bread, stared, marveling, no doubt, at the wild ways of the foreigners. Some of the boys had picked up unoccupied females and were using the barrack rooms as they would a dirty wineshop, caring naught for the audience of young and old.

And right in the middle of this crazy night of debauchery, this temporary cessation of one of the bloodiest battles in the history of the Legion special medals were struck for those of us who survived the murderer of our four comrades was discovered, and in the simplest manner. McCann and I were among the first to hear of it. We were suddenly shocked out of our drunken carousal by the appearance of a young Swede, whose friendship we had recently cultivated. He was as wild-eyed and excited as the devil. When at last he was able to reach us through the wine fumes, we understood that he had seen personal property belonging to some of the murdered men watches and some looted antiques that they had been known to possess.

"You're drunk, you Swedish runt! Where'n hell you see all this stuff?" growled McCann.

"You come, Englander! I show! Quick!"

"Come on, Mac! There may be something in this!"

We followed Jacks through the barrack rooms filled with drunken men, women and children, to a corner of the drill yard, where we found a man and a woman lying together, dead drunk. The man was a Russian belonging to our Squadron, the woman was a so called Christian Arab from the town, and she was literally smothered in loot; bracelets, necklaces a girdle crusted with stones and stuck through with a jeweled knife, a number of rings, and a sort of sirpesh made of gold and silver coins. One thing was certain, she had never entered the Citadel with all that stuff about her person.

The Russian, when he saw what we were about, tried to put up a fight, but he was too far gone in liquor, and we soon had him down. We stripped both the woman and him of the loot. Two of the watches we found on him were inscribed with the names of the dead legionnaires. It soon became obvious to us that the fellow must have known of the valuable loot those four men possessed, and he had made up his mind to get it. We took him along to the Skipper, and reported the find.

Captain Dupont himself was far from being sober, but it did not take him long to get the drift of the story. It took him about three minutes to make his decision. The Russian was propped up against the fort wall there and then, six men were turned out, and phut! That was the end of the murderer. And it all began through the big Russian stealing off with the little Swede's girl! He had followed and watched. Had seen the drunken display of wealth, the maudlin lovemaking and the decoration of the Arabian woman. It was when he saw the display of watches that young Jacks, from his hiding place, began to feel something more than a lover's jealousy. He became suddenly suspicious . . . and had then sought Mac and myself with the story.

After that we went back to the carousing crew in the barracks, and fell in where we had left off, so to speak We had wasted enough energy on a murderer, and it seemed that we tried to make up for lost time. Opportunities to do so lay thickly around us.

Dawn and a hail of bullets roused us from our stupor. The Skipper, his officers and N.C.O.'s, raced round and used their whips very freely. The men crowded to the walls, started the machine guns, and yelled at the enemy to come and take us if they could. Well, it looked as if they could! Sultan-el-Atrash's mighty hordes had been greatly reinforced during the night, and they crowded up towards the Citadel with spears, swords and rifles, as if they meant to take us in one headlong charge.

Why they did not take the Citadel in that first mad rush will ever remain a mystery. They actually reached the big gates, and while some of the Druses poured petrol over the timbered structure, numbers of others crowded about them to ward off our withering fire. Scores of them went down in that crazy attempt to fire the gates, but they succeeded in getting them alight, and soon the flames were soaring, scorching the men who fought on the wall above to protect them.

Lieutenant Casteg and six of our men volunteered to go out and quench the flames. They went at the peril of their lives, and before their work was done had sacrificed their lives, for not one of them was left standing despite our efforts to mow down their attackers. The bodies of dead Druses were piled up under the fort walls and used as steps. One section of the walls was actually taken, and the attackers were using this vantage to pick off our men.

The sight of those Druses on the wall was the signal for renewed yelling and screaming on the part of our women and children. The Skipper, plastered with mud and sweat, like his little band of men, was raving and cursing, calling upon his dogs and pigs to exert themselves. It did not seem to matter how he exposed himself, for he went through that ghastly day with only a grazed arm. Suddenly the enemy withdrew, and began to race back down the road as if chased by the whole French Army.

That could mean only one thing. Our relief had been sighted. Away in the distance we saw an advancing cloud of dust. Nearer and nearer it came, until at last we could make out the various regiments that made up a vast column advancing at the gallop. It was Colonel Andre with two squadrons of Spahis Tunisian, a number of companies of Algerian troops, and a bunch of the 27th Colonials. We learned that a mounted party of Christian Arabs had escaped from the town as soon as the trouble started, and had made their way to Damascus, where they had informed the Commandant of the gathering of Druses in the hills and the advance on Rashaya.

Colonel Andre found us a small, battle-scarred band of legionnaires and a handful of Senegalese. And he was pleased to decorate us with the Flag of France.

"France is proud of you, my legionnaires," he said.

And we grinned, thinking this a good joke.

All we were concerned about was getting back to Damascus and something resembling civilization. We were becoming rather satiated with French blunders.

CHAPTER XIII: MAKING REBELS AND DESERTERS

When we had last visited Damascus the city was in a state of siege, and it had been necessary for us to fight our way through the streets to the Citadel. On our return from Rashaya we found it quiet, ominously quiet. There was a brooding atmosphere over the place which boded ill for someone. Not only had order been completely restored, but the authorities were ruling with an iron hand. There were to be no more revolutionary outbursts. The slightest sign of disorder on the part of the inhabitants, and blood flowed. If one were a native of that colorful city one walked the streets warily.

The citizens were not merely subdued. They were cowed. The reason for which was not far to seek. We had not been in the place many hours when we realized that a wholesale system of flogging was in progress. It is not my intention to describe those floggings. They were too many and too hideous. But it is my intention, at this stage, to make a specific indictment against those who took command of Damascus immediately after that historic siege.

It seemed that something had to be done about the official reports of the siege. It was necessary that the French should come out well on the credit side, that there should be some justification shown for the appalling bloodshed of the period. Possibly other nations were interested in the affair, certainly the League of Nations would be.

Now, with regard to a rebellion of this sort, it is not enough to state that it took place. It has to be proved. And what is of equal importance, it is essential to find the cause. One way of doing this is to gather into the archives of authority as many confessions of guilt from rebels as possible. On the face of it, not a very easy task.

The French, however, have ways of their own in attending to these little matters. There came, therefore, a bloodthirsty period when natives of the city, be they Druses, Jews, Bedouins, Christian Arabs, Syrians, or mere fellahin, were seized and thrown into prison on the slightest pretext. If, for example, a citizen at that time were to object to the way the troops manhandled his wife, his daughter, or his sister, he was arrested for attempting to revolt against law and order. This was indeed a common method of procuring rebels. It became so common in some quarters of the city that the women, rather than have their menfolk carried off to jail, never to return, offered themselves to the patrols. . . .

Nevertheless, vast numbers of innocent natives were thrown into Damascus jail, and once they were inside—God help them! These prisoners were asked to sign statements, written in Arabic, the language of all creeds and classes in Damascus, which made them self-confessed rebels. If an unfortunate native refused, saying he was a peaceful citizen and desired to remain at peace with the French authorities, he was flogged into insensibility. If, on recovery, he again refused, he was again flogged until unconscious. In some cases, respectable, peace-loving citizens were subjected to three or four such floggings before they finally signed the false documents.

Having signed such a confession, the prisoner released when it was too late that he had put his hand to his own death warrant. As a self-confessed rebel he was placed up against a wall with others and shot 1 In the eyes of the world, the French were justified in putting rebels to death. AU they required was proof that a rebel was a rebel. Could anything be simpler? Could anything be more uncivilized? Is it for this that the peoples of the western world have been progressing for twenty centuries?

I know nothing of the official reports regarding the Siege of Damascus and its aftermath. It is conceivable that their accounts of captured rebels do not coincide with mine. I am giving my own experiences. I am telling of what passed before my eyes. No amount of official contradiction, or national spleen, shall destroy the evidence of my own sight. I saw what I saw. I am stating facts. Even though I were a legionnaire, brutal, vile, and as disgustingly lascivious as the rest of my comrades, I could still think and feel as an Englishman. I was never so infused with the lust for blood and women as not to know what I was doing. As a legionnaire, I lived the life of a legionnaire, obviously. A legionnaire is a loathsome brute. If he is not, he dies. There is no other alternative.

And yet, there were legionnaires, vile and bestial legionnaires, who sickened at the perpetual shooting parades. It was one thing to kill men while in action of battle. That was self-defense. But it was quite another to shoot down innocent citizens in cold blood. I have killed a good many men, and derived a kind of satisfaction from the killings. It was they or me. What would you? I have been placed up against a wall to be shot accused of killing twenty-four Partisans. The French authorities called me murderer. I know that I acted in self-defense. I fought for my life. But that is a matter for a later chapter of this chronicle.

What filled me with a sickening nausea was to find myself in those damnable firing squads, shooting down innocent citizens, some of whom had been flogged, twice, thrice, even four times those self-confessed rebels thrashed into providing evidence for a League of Nations. Even a legionnaire has limits. I was not alone in this peculiar feeling. McCann, another crack shot, young Jacks, another, Hoffman, another, several close friends of mine at the time, gave expression to it. Sufficient evidence in this, I think, to show to what extent French authority could go when they would sicken even legionnaires.

Those hideous repetitions not only sickened a great many men of the Legion, there was a natural reaction among them. Used as they were to doing the dirty work of the French Army, the day by day parading in firing squads proved the last straw. This making of rebels to be shot also made deserters. Quite a big number got away from Damascus at that time. What happened to them, God alone knows No doubt the Partisans accounted for some of them.

And the Druses accounted for others, including some of my own friends. The Druses had a way of gathering information about the affairs of the Legion. They certainly got to know about the revolt against firing squad duty, and they took advantage of the opportunity to induce legionnaires to desert and go over to the enemy. In some cases, big sums were offered. McCann and I heard of this through our friend, Hoffman.

Hoffman, it should be explained, was one of those fellows who were perpetually hard-up. In spite of his opportunities and the cheapness of the wine, he never seemed to be able to buy his own. Otherwise, he was a good fellow. He was an ex-officer of the German Army. His was the old, old story of making love to another officer's wife, which ended in the death of the other officer, and Hoffman's flight and enlistment in the Legion.

To our amazement, he rolled up at the *Cafe Noir* one evening with his pockets bursting with cash and insisted on paying for everyone's drink.

"Hey, Hoffman! Where in hell d'ja get all this dough? You been on a private looting expedition?" queried McCann.

Hoffman drew us aside with the air of one who was about to reveal State secrets. I suppose they were State secrets, for his information was such as to affect the State of Damascus at that time.

"My friends," he whispered portentously, "I know I can trust you both. I am going to tell you a big secret. But, first, I must ask you a question. How would you like to desert and become rich men?" "Haw, don't be crazy," growled McCann. "The hooch has gone to your head."

"Shut up, Mac!" I said. "What's the idea, Hoffman? Is this a joke of yours? "

"I do not joke, Barrington. Listen! To-morrow night I shall desert. I am joining the bandits, and I shall live in the Lebanon mountains henceforth. I have met a Syrian girl. .

"Ah!" smiled McCann, "nice cup o' tea?"

"No, no! Not so! She is what you would call a fine girl. Her father is a big chief among these Druse bandits. He has been travelling in Europe and has a gang of gun-runners serving him. Already he has 250 brand new rifles. He is very rich. He wants a number of us fellows from the Legion to help drill his men. He is raising an army. He will drive out the French one day."

"Say, boy!" laughed McCann, "what a goldarned ambition that is! Who wants to drive out the French, anyway? They can have the damned country, so far as I am concerned. And how d'ja know this old joker is rich, eh?"

"I do not he. Look!"

Hoffman bent over the table and displayed for our benefit some five thousand francs.

"There is plenty more! And soon I shall be at the head of thousands of well-drilled men. Now, my friends, what will you do? Come! Leave this filthy rotten Legion that pays you only a few sous a day. There are five others coming with me. I will not tell you who they are. You will know soon enough."

"Hell, Barrington! I'm not saying this doesn't appeal to me."

I shook my head, telling McCann to go if he wished, though, heaven knows, it would not be my wish. Nevertheless, I should desert myself when the time came. But not in this way. I had not been idle, for I had learned that Nashalla had left Homs. Where she was I could not discover, but I was only waiting for a sign. In any case, I should not take such a way out. It meant becoming an outlaw. Not that I was afraid of being an outlaw. But it led nowhere, and I said as much.

"No, Hoffman. You can count me out. I don't trust these bandits of Druses. They are worse than the Sultan-el-Atrash's army, always squabbling among themselves. There'll be trouble over the loot and the women and that sort of thing.

I know the game."

He shrugged his shoulders and stared at McCann.

"Waal, Barrington, you and me have had some tough breaks together. I guess I stand by what you say, boy. We'll cut it out this trip. Yeah?"

Hoffman was visibly disappointed. Apparently he had counted on us, knowing us to be the Skipper's own trained crack shots.

"Quit worrying, Hoffman. You know your little get-away is safe with us."

"I know, I know, friend McCann. It is not of that I am worrying. I had hoped you would come. Well," with another characteristic shrug, "it cannot be."

McCann laughed suddenly.

"Say, Hoffman! Have you thought what this might mean, eh? D'ja know we may come on you in the hills one day, and slaughter the lot of you!"

I must confess I had not looked at it in that light, but Hoffman only smiled in his fatalistic way.

"To-day we are very good friends,'" he said. "Another day, maybe, we shall cut each other's throats. Who can say? It is what you call, all in life! Once the German fought the English and the American. Now they fight side by side in the Legion. What is the difference, my friends? "

"Waal, if you want to look at it like that there ain't any difference. A man fights, gets his rations, and his loot, and his nice cup o' tea now and again. You've spilled it, Hoffman. There ain't no difference!" "Well, my friends, I am very disappointed that you do not come. I like you both. I would rather have you two behind me in a fight than anyone I know . . . "

"Haw! Cut it out, boy, cut it out! Kim on! Let's have some hooch!"

We drank—to our next meeting!

Three very drunken legionnaires returned to barracks that night, one of them to spend his last night with the Legion. As he said, it was all in life. Apparently he had gone very deeply into the task that lay before him, for next morning he handed to me a slip of paper on which he had written, and transcribed into English, the battle chant of the Druses.

"Upon them! Upon them! Oh, Lord our God, That the foe may fall in swathes before our swords.

Upon them! Upon them! That our spears may drink at their hearts.

Let the babe leave its mother's breast, Let

the young man arise and be gone!

Upon them! Upon them! Oh, Lord our God,

That our swords may drink at their hearts! "

All considered, it seemed a particularly blood-thirsty sort of chant. But, after all, was there much difference between that chant and the hymns and prayers of other nationalities during war?

The following day another bunch of legionnaires had gone. Hoffman and his merry men had deserted. The usual patrol was dispatched in search of them, but they returned with nothing to report.

"*I* almost wish I'd gone with the bastards," growled McCann. "Oh, boy, just fancy you and me leading a bunch of lousy Arabs in a charge against the Legion!" Tcha! He spat angrily.

"How in hell can we fancy any such thing! Gosh! I'd like to see that mob facing the Skipper when he's well primed, eh? Barrington, you son of a gun, you were dead right. That wasn't our trip. Kim on! Let's put a few more rebels away. Wonder how long this goldamed game is to be carried on. Gee, boy, the sight of those poor devils . . . stuck up there. . . . Kim on! I'm getting too blasted sentimental."

CHAPTER XIV: MASSACRE OF MEZZRA

But, after all, I can offer no more damning indictment against France than this—the appalling massacre in the Mezzra Bowl. It was as if some giant hand had picked up a whole army of men and guns, equipment, accoutrement, tanks, transport and beasts, and dropped them all pell-mell into a seething cauldron. I doubt if there has ever been anything quite like this astounding blunder in all military history.

Blunders, of course, are inseparable from military activities. It seems to be a question of what can be foreseen and what cannot by the average intelligence. My view of this terrible disaster is that it ought to have been foreseen, that, indeed, it *was* foreseen by those who had no power of rank to prevent it. Men, very ordinary, every-day sort of men, with average intelligences and below, were amazed at the maneuver, but with the stoicism of their kind, shrugged their shoulders and obeyed.

Mezzra lies in a huge bowl almost entirely surrounded by mountains. It is commonly called the Mezzra Bowl, and lies some little distance below the town of Sueida, the stronghold of the Druses. Naturally, it is well watched by the Druses, since by its very nature it makes an admirable trap. Which is another reason why this bloody massacre ought never to have taken place. We of the Legion cavalry, the indomitable Fifth, headed a column marching on Sueida under command of Colonel Andre. The column consisted of Legion infantry, four French batteries of 75's, three battalions of Colonials, four squadrons of French cavalry, a tank corps, and transport. A goodly crowd to move about anywhere, let alone the mountainous district of the Jebel Druses.

Our Squadron, spread out, was the dredger and guard for the advancing column, and we marched a day ahead of the rest. As usual, we were apportioned the job of heading off any trouble that might happen along, and so leave the way clear for the advancing column. In this particular case, however, our position was the means of saving us from being cut to bits when the avalanche came.

We reached the Mezzra Bowl and would have passed on, but were called back by Colonel Andre. *He had decided that the column should camp in the Bowl for the night!* But for their protection the

Fifth was to turn back and cover a village in the rear. That command was the most fortunate that had ever been given to the Fifth! In effect, it ordered us out of the Bowl. We were to cover the rear flanks. Why, nobody knew.

The mission of the column was the relief of a small French force that was being besieged in Sueida. Mezzra is some five miles or so below Sueida. Yet, instead of keeping our position in advance of the column, we were turned back to cover a village in the rear while the column camped for a night in the Mezzra Bowl! It was, to say the least, an extraordinary situation!

The camp was quiet when we left it behind, but we had hardly got in sight of the village in question when the terrifying avalanche descended. Literally out of the blue of the mountain mists came 20,000 Druses. Imagine the confusion, the amazing pandemonium that ensued! Of what earthly use were tanks and big guns? Men were stabbed to death before they could swing the guns. The column was trapped as no column had ever been trapped before. It was child's play for the Druses. They simply love hand-to-hand fighting, and in that respect they had the time of their lives! Several hundred men were slain before they had time to shake the sleep from their eyes. Horses stampeded, Commandants were yelling orders in all directions, utterly useless and imbecile commands that no man could hear and could not obey had he heard. The Druses from the first moment took possession of the column and screamed their way in and about the encampment in a passion of wholesale butchery. In an incredibly short space of time there was no fight left in the column, dead and wounded were lying about in heaps, the Druses were kicking up an infernal din in their destruction of the tanks they had captured, since they had no use for such engines of war.

It was the finest haul the Druses had ever made equipment by the ton, thousands of rounds of ammunition, scores of machine-guns, rifles, Mauser pistols, sabers, valuable transport, rations for a column, horses and men! No wonder they were jubilant! At last they would drive the French to the sea. They certainly drove the remnants of that column. We arrived on the scene to find Colonel Andre and his staff on the run, with the Legion infantry endeavoring to cover the retreat.

We were promptly turned into the scrimmage. What a slaughterhouse it was! The Druses, drunk with victory, were driving the rearguard back and back, hundreds of men and horses lay scattered about the Bowl, wounded and dying were being tortured until they screamed in agony. The Druses were simply wallowing in a feast of blood. We threw our weight into the rearguard, and contrived to trample a number of the maniacs down before we, too, must turn and make a dash for it.

So we fought and retreated, just as we had done on the retreat to Fez, but this time the Druses did not keep it up so long. They were more concerned about their captures in the Bowl, and though the whole body of them did not come out in the open to hasten our retreat, they came in sufficient numbers to cut up our Squadron pretty badly and practically to annihilate the Legion infantry. "Hey! You sons o' bitches! We're coming back!" cried McCann, who could make as much noise as any Druse, "we're coming back to-morrow!"

It was certainly astounding that he could make himself heard above that ear-splitting din. The fiendish yelling did not cease until they had driven us several miles back from Mezzra, and at last retired to join the remainder of their vast army in the Bowl. Nearly three thousand men were sacrificed in that massacre. That was not all. Another column of French, Colonial and Legion troops was cut up in an attempt to relieve Sueida though my Squadron did not take part in this, and I cannot, therefore, give an account of what took place.

One thing I do know, there was only one Massacre of Mezzra!

CHAPTER XV: FOUR WERE TATTOOED

After Mezzra Gazalie, of all places, where we indulged in a so-called rest. A rest for the Legion usually meant a perpetual round of all manner of disagreeable duties, such as tax collecting, chasing bands of brigands, burying other people's dead, “cleaning up" some troublesome quarter, breaking in mules and horses, and so forth. An ordinary day began at five a.m., with horse parade fifteen minutes later, five hours exercising the animals, and then the day proper, which was filled with duties and drills until six or seven in the evening.

In this region the country everywhere was beautifully wooded and very fertile. Our encampment was on an eminence overlooking a scattered group of villages. There appeared to be an abundance of gardens, orchards, and fresh green life generally. If one had been a territorial instead of a legionnaire, it might have been an ideal spot in which to camp. But the Legion contaminated all.

Its chief appeal was that it made a step nearer the Palestine Border and offered excellent cover for deserters. Therefore, we had a great many fellows making the attempt, and for those who did not there was always patrol duty. At that time about a score or so of my comrades approached me for information about Palestine. Being a Britisher, I was sure to know all about it. Palestine was the Mecca of the legionnaire, and, apparently, I could assure him of a square deal once he had reached the coveted land. There were so many ill-starred attempts! It was pitiful to see the poor wretches dragged back, either by the Partisans or by our own patrols.

On one occasion eleven men cleared off in a bunch. The discovery was made at about three in the morning and an hour later the Skipper himself, with a patrol of thirty men, set off in pursuit. His blood was up. He was determined to find these men if he had to follow them all the way to the Border. We combed the surrounding woods for hours without coming upon a sign of them. Then we began to take in the few villages that lay directly between our camp and the open country, and as we did so we realized how difficult a job we had set ourselves.

"Gosh, boy! This is the best place we’ve struck yet for a getaway. What say, Barrington?”

"I like it, Mac, I like it. It's ideal. While we're through one village, the fellows can be in another, while we're doing one bunch of scrub, they can be clearing another. There's nothing in Africa or Syria to touch it. Mac, it was made for deserting legionnaires."

"Shoor, boy, you've said it. Now, what about it? I've got enough loot to keep me the next ten years, and so've you. What say? " I shook my head, and smiled.

"I'd love to—but."

"But nothing, you big sap! You getting timid? "

We were approaching a village. In another minute or so we should dismount and make a house-to-house search.

"Timid yourself, you big stiff! I've told you. I'm going to find Nashalla before I quit. Then, if you like, you can join us."

"Thanks for nothing! You must be crazy! Where'n hell you going to find this cup o' tea"

"She's not! I think I know a decent girl when I see one. I've seen enough of the other kind. Can that stuff about cups of tea. I'm telling you, McCann, she's different."

"Who is she? What is she? All this damned palaver about the bitch . .

But further argument was cut short by the order to dismount. We started to search the houses and courtyards. It was a job after our own hearts, there being plenty of opportunities for loot and other things. The inhabitants of the village were scared stiff at the sight of the Legion patrol, for they were mostly unintelligent Arabs of the nomadic order, and perfectly harmless.

Our procedure was to kick open a door and enter without ceremony. Usually we would find an old Arab with his bunch of women and children, perhaps in the middle of a meal, or industriously working. If the householder protested he was thrown into the street and told to remain there until we gave him permission to return. If he were perfectly civil and his women happened to be more or less attractive, he was still thrown into the street. Sometimes those helpless women made a noise. More often they would shrug their shoulders and take what Fate sent them. Some of the older and more experienced ones would dismiss their children as soon as we appeared, throw themselves on the floor and cover their heads with their clothing. In the main, it was the young girls who put up a fight. When we heard piteous childish cries coming from a house we knew that some girl in her early teens was trying to keep her virginity. Those tremulous screams. God! I can hear them now! They were enough to soften the heart of any man, except a lecherous legionnaire. . . .

McCann, bad as he was, would suddenly pale under the tan of his India-rubber face, like a man about to be sick.

"Gosh, Barrington, why'n hell don' they leave the kids alone!"

Loot was another matter. We could loot anywhere. McCann had a nose for loot. He could smell it yards away, and ferret it out however well it were hidden. He could also tell at a glance if a thing were mere rubbish or whether it would have value in the market. At this time we must have had about seven thousand pounds' worth between us, and it was becoming a sore trial. We had to get up to all sorts of subterfuge in safeguarding the stuff. Quite apart from the danger of having some officer or N.C.O. discover it when they would immediately confiscate it there were covetous eyes among one's comrades. No legionnaire would trust another where loot or women were concerned, unless, of course, the two were fellow countrymen. That was different. One's word of honor might mean something in such a case.

"Curiomania" had full sway on that particular day, for we spent the better part of the morning in the nearby villages. At noon we headed for the open country, still, ostensibly, searching for the eleven deserters. Hour after hour we continued, at a walk-march and in extended order, to sweep the plains, drowsing in our saddles under the terrific heat of the sun, hoping in our hearts that we should not come upon the unfortunate crew. And, just as there seemed a possibility of our making an about-turn, there came into view on the distant horizon what looked like a native encampment.

The order was given to proceed at the gallop and we raced towards the camp. In less than an hour the sun would be gone and we should, likely enough, spend half the night in getting back to our own quarters. We came upon the native encampment in a cloud of dust, wondering greatly if this would prove to be the refuge of the absentees. It proved to be the halt of a Bedouin tribe, a cluster of a dozen rude dwellings of reeds and canvas thrown up in the shelter of some date palms. In the center burned a brazier, on which was set a huge iron vessel. It gave oft a very appetizing smell. The Bedouins who had been seated around it jumped to their feet as we rode up.

The Skipper rapped out an order and the patrol moved in a wide circle about the camp and halted. If the deserters had found this spot they could now abandon all hope of getting away, for they were hemmed in. We made inquiries of the group of natives. There must have been about fifty of them, men, women and children. They had not seen nor heard anything of the deserters. They had been camped at the oasis for two days and they must certainly have seen the band of soldiers, had they passed that way. As we could not think of taking the word of a native, several men dismounted and a thorough search was made of the crazy hutments, but without result.

The natives were friendly, made us welcome of their humble abodes, and offered us food. To our relief, for we had been on the move since four o'clock in the morning, the Skipper decided to camp for the night. Guards were mounted and the remainder of us, having tethered and fed the horses, were at liberty to make our beds or wander about as we pleased. The Skipper had taken possession of one of the huts, and when I went in to attend to him I was quickly booted out again! He would not require my attentions for the next few hours.

We were up again at daybreak. The night had been uneventful, and there had been no movement over the country to indicate the presence of wandering legionnaires. We rode steadily back towards camp, searching the countryside as we went along.

Nothing happened to break the monotony of the day's march until late afternoon, when we were within sight of the villages that marked the fringe of our encampment, and were proceeding *au trote* along the bank of a creek. The Skipper's glasses had picked out some objects in the water ahead of us. When we drew level with them we found four of the wretched deserters vigorously trying to make the opposite bank. Captain Dupont's Mauser spat once, and one of the heads disappeared into the water. We thought he had registered a hit, but presently the head bobbed up again. The swimmer had anticipated the shot and had ducked under, only to come up again for air. The order was given to fire. We sat our mounts on the banks of that stream and took potshots at the four bobbing heads. It was a ghastly game, a sort of gruesome aunt-sally business!

I have had some queer targets in the course of a not uneventful life, in and out of armies, but never any quite like those. The four heads streaming with water would emerge for a few seconds, shots would ring out, and down they would go, staying under as long as it was humanly possible even for good swimmers. Then one head came up streaming with blood as well as water, there were red ripples on the surface, arms and shoulders shot up and splashed under, never to rise again. The other three turned about and signaled that they would surrender. We lowered our guns and waited.

What a sight they were—those three miserable legionnaires! They wore only trousers and shirts, which, of course, were drenched. They trod the water wearily, and slouched up to Captain Dupont like men utterly spent, as indeed they were. Such a miserable sight ought to have excited pity. It didn't. The Skipper bent forward in his saddle, took one long look at the trio, as if he could not believe the evidence of his own sight, and let out a terrific roar of laughter. This was so unexpected to us behind that we could only stare in amazement.

"Mon Dieu!" he cried, "but these fools have been dealt with already. Look! Are they not branded?" And he roared again.

He gave the order for them to be tied up to our saddles by the neck, and as we approached them to carry out this fiendish command we saw the cause of his laughter. Somewhere, somehow, the foreheads of those three poor devils had been tattooed with lewd figures. How it had been done, and why, we learned later, but in those moments, when we could not take our eyes from the nauseatingly clever outlines that had been pricked into the skins of their brows, we did not know whether to giggle or curse. At least that was my personal feeling. On each brow was stabbed indelibly the figure of a nude woman, postured in a fashion typical of a certain Eastern mentality. They had obviously

tried to scratch out the tattooist's handiwork, but had only succeeded in lacerating the skin. The filthy black outline remained for all the world to see . . . and laugh at! God! Were they to go about for the rest of their lives with faces covered because of that unsightliness, that beastly obscenity!

When they had been secured to the saddles, the Skipper, between bursts of laughter, began to question them, inquiring how the tattooing had been done. They confessed that four of them had broken away from the other deserters and had sought refuge in the house of a native. It was their intention to remain under cover during the day. They had given some loot to the native for his hospitality. He had fed them and they had drunk of his coffee which was drugged. They had awakened to find themselves miles away from the village, down that very creek, in fact, all their money and loot gone, also most of their clothing, equipment, water-bottles, everything except what they stood up in trousers and shirt. And then they had seen the sickening tattooing on the faces of each other. . . .

And all the time Captain Dupont roared with merriment at the pitiful wretches, becoming unspeakably obscene in his references to the lewdly pictorial foreheads. When he grew tired of torturing his victims, the order was given to proceed towards the village at a gallop. The trio were to direct us to the house where this atrocity had been committed, not because the Skipper was anxious to avenge this wrong done to his men, but so that he might see what manner of personality was he who could devise such an atrocious performance.

So the drenched, worn and utterly exhausted caricatures of men were dragged along by the neck at the gallop, mile after mile, until they yelled their agony. Only once did we halt. It was when I protested that my prisoner was dead and that to drag the body along the ground in that fashion merely impeded my horse's movement. I drew the drenched and mud-caked body up into the saddle, turned it over so that I could not see the frightful picture, and away we went once more.

On arrival at the village we dropped into a walk-march the two prisoners directing us towards the house we sought. We must have presented a weird sight to the natives of that village, a patrol of thirty men, covered in dust and sweat, leading two wretched prisoners by the neck who looked as if they had been dragged through a bed of mud. The Skipper rode ahead with his whip twirling through the air, ready to flick any too-curious native who dared to stand and stare.

A shout from our prisoners brought us to a halt before a house. We dismounted and strode up to the door. It was locked. We hammered for a few minutes without getting an answer. Finally we broke down the door and stepped into the courtyard. There was not a soul about. We raked the place from top to bottom. It was empty. The tattoo artist had flown. We gathered the few sticks of furniture together and set fire to the house. It was blazing merrily as we turned our backs upon it and returned to camp.

The problem of those two prisoners, as to what they should do with their hideous faces, was solved two days later. They were shot for desertion.

And yet there were men in that patrol, seeing all that they had seen with regard to attempts at desertion, who themselves deserted later finally succeeding in escaping!

CHAPTER XVI: SUEIDA—KINGDOM OF THE DRUSES

We joined in the third and final attempt to relieve Sueida, the capital of the Jebel Druses, in which a small French force had been living in a state of siege for weeks. Sultan- el-At rash's mighty hordes, now crazy with delight over their colossal victory at Mezzra, had laid siege to the villages and hamlets for many miles around, and the whole of Southern Syria was in a veritable uproar.

This, of course, was the Jebel Druses' own country, backed by the range of mountains from which the tribes gained their name. It was obvious that they would fight for what they had gained of the plains and open country, and doubtless they dreamed of a new and bigger kingdom that was to grow out of the fertile plains.

We of the Fifth, however, were heading a vastly different column this time. The French were taking no more long chances in this bloodthirsty problem of Sueida. General Andre's column was reinforced to full strength, and twenty bombing 'planes escorted us out from Gazalie.

Four days' march lay ahead of us, and through country seething with fighters drunk with success. This time we took a different route, avoiding Mezzra as we would the plague, proceeding with that Bowl of Death on our left flank and Bosra on our right. The Legion cavalry, marching a day ahead of the column in extended order, was escorted by three of the 'planes. The fun began as soon as we reached the open country. The enemy was spread around everywhere like flies and for three and a half of those four days we rode through a network of snipers and harassing detachments. All over the desert and plain they had been busy building breastworks, small crude stone walls, from behind which they kept up an incessant sniping. The 'planes bombed them from the air, and we were continually making charges.

Before we had left Gazalie, General Andre had addressed our Squadron, saying: "Men of the Legion cavalry, I will give you the honor of entering Sueida first!"

And as we rode and fought our way through the infested regions that lay between Gazalie and Sueida, those momentous words of the brave General stuck in our gills. We might enter Sueida first, but, in all conscience, we had to get there first. Druses are excellent marksmen. They are about the finest snipers in the world. They not only brought down men and horses, they showed their marksmanship by wrecking one of the 'planes. They had hundreds of those stone walls dotted about the countryside. It was their own country. They knew it well. They fortified nullahs and mounds. Bombing 'planes might bomb them. They would disappear underground, and we had to ferret them out before we could proceed. Altogether a lively time.

Sueida, an ancient walled city, comprising the headquarters of Sultan-el-Atrash himself, an ancient burial ground, and the Temple where once was practiced the Worship of the Golden Calf, holds a most strategic position, being built on the crest of a hill overlooking vast tracks of desert, and is protected from behind by mountains.

The column joined us at the foot of this approach to the famous city, and the French 75's opened fire on it, bombarding the place for six solid hours and so short-ranging at times that they killed nearly two hundred of our men in the front flanks of the column. Between them and the bombing 'planes, the greater part of the city walls and many of the buildings, including the Citadel, were brought down in ruins. When the Fifth Squadron entered the city first, as promised by General Andre to sweep up the remains, there was little left to fight. There were a number of dead and wounded lying about the roads, but the vast majority of the inhabitants and the great Druse fighting forces had fled to the mountain fastnesses behind the city.

We had taken Sueida, but the trouble was by no means over. We had not conquered the great army of Druses. We had merely driven them, with their women and families and goods and chattels, into the mountains that overhung the city, into the caves and fastnesses which they knew so well. Our airplanes soared over the region and bombed the mountain sides incessantly, but for all the damage they did they might just as well have dropped bombs into the sand of the Syrian desert.

When we had given the all-clear to the column waiting without, General Andre gave the order to advance, and the vast column marched into the city with all the pomp and ceremony of a victorious army, banners flying and bands playing, tumultuous cheering *" Vive la France! Vive la France!"*

General Andre then brought out his box of medals and duly decorated us.

Down on the plain below us lay our dead comrades, Spahis Moroccan and Colonials as well as men of the Legion.

We were made free of the city, or what was left of it, and the men of the Fifth proceeded to get roaring drunk, much to the disgust of some of the French troops.

"Oh, boy!" roared McCann, "this isn't a war. It's a picnic! Think of all the troops, thousands of 'em around, *and* an ambulance! An ambulance, you big sap! An ambulance! "

True, it was a great little battle for the Legion, that entry into Sueida. It was something fresh in warfare for us. We had usually been in the position of the poor devils who had tried to hold their own in Sueida the defenders, not the attackers.

Next day we got back to normal, beginning our duties by burying the dead. It was a gruesome business that had its compensations. We were able to replenish our kit at leisure here a good pair of top boots, there a fine Mauser pistol, or a knife, or even personal property, to say nothing of stray loot. Having cleared away the dead, we were set to re-building the walls and cleaning up the roads.

Day after day the Druses kept up an incessant sniping from the mountains behind us, picking off a horse here and a man there. It never stopped. Whatever we were doing, watering the horses, standing on parade, making roads, carrying rations, the sniping went on phut, phut, phut! The 'planes hovered above the mountains and bombed, but all to no purpose. They could not reach the secret caves of the Druses.

And while the sniping continued, other chiefs of the Druses came into the city to sign their pledges of peace and acknowledge the French Flag. On the plain immediately below us, where Spahis Moroccan patrolled, guarding the cattle that the column had collected *en route,* the snipers were busy, harassing and menacing the camp.

They were also at work further afield, for our Squadron was turned out to go in search of a transport column that was unable to make Sueida on account of the marauding bands. We found the transport, after a couple of days, with some of their wagons wrecked, horses broken loose, and the men fighting for their lives. Having driven off the bandits, we repaired the damage as best we could, and swept the surrounding country in a series of charges before the transport finally reached the walls of Sueida.

That, however, was but the beginning of a series of skirmishes marked down for the Legion cavalry. We found ourselves dedicated to the job of hunting bandits and stray detachments of Druses who infested the mountains behind Sueida. The mountain passes proved to be treacherous country for us, for the bandits had the advantage. They knew the ground, every hole and cranny of it. We, on the other hand, must, perforce, move blindly, never knowing what we should find round the next corner. Also, it was difficult going for the mounts.

We endured this guerilla warfare for about ten days, and lost four officers and thirty men in the process. Never a day, never for a single hour, during our sojourn in Sueida, were we permitted any respite. Our Squadron developed a definitely mutinous temper. There was the everlasting talk of desertion, though how anyone could hope to get away from such a place I could not imagine then, yet I came to do more than imagine it a fortnight later. I actually attempted it!

There were two things that brought about our change of mind with regard to the possibility of escaping from the Legion. One was the news that, after we had done all the dirty work necessary in Sueida, which might take several weeks, we were to be passed on without rest to the outpost of Derra-Zor, a spot rumored to be Syria's most perfect hell-hole. I never went there to confirm the rumor. The other incident occurred during a pitch battle in the mountains.

We were approaching a narrow gulley when an Arab spy came up and reported to the Skipper that a large band was moving towards us. We immediately dropped back and sought concealment behind a head of rocks. From this point of vantage we watched them approach. There must have been two or three hundred of them, mounted, carrying all sorts of equipment, modem rifles and sabers as well as old-fashioned guns and long straight swords. Some of them wore chest plates that glistened in the sun, and all carried dozens of rounds of ammunition on their persons and on the horses. Behind them rode their women, bringing with them the transport wagons. Verily, this was a new army, extremely well equipped and accoutered for battle. And surely they would get the battle. We watched their advance with something like amusement. They looked for all the world like an army of days long gone by, marching straight out of a child's picture book.

They came on, marvelously sure of themselves, and totally unconscious of what was awaiting them. We knew what it was like to be trapped. We had experienced the sensation on three or four occasions, though we had never met a force quite so big as this since we started to scour the mountain recesses. They came almost on top of us before the order to charge was given.

We dashed out upon them like a lot of fiends lusting for blood, which perhaps we were. God! What a frightful scramble it was! We were in their midst, slashing right and left before they had time to draw. No waiting to go over the top here! Just old fashioned, blood-spilling, hand- to-hand fighting. Many of them were thrown from their horses and trampled into the rock-strewn road by the fierce onrush. The women behind set up a terrific screaming, while the fighters on both sides cursed and yelled at each other. In the middle of it all, swinging a saber as no other man could, was Captain Dupont, he of the charmed life, having it all his own way in this, his true element. In less than half-an hour we had hacked our way through the bunch of them, scores of them beheaded or with limbs severed lay about that narrow defile, and we charged on after the fleeing rearguard

We rounded up the sixty or so who remained on their feet and took them prisoners. Some of the Squadron had chased after the women, and returned to the head of the gulley with most of them. They were dumped into their own wagons and placed under guard. What a gory scene it was! What a picture for those shivering camp followers to gaze upon! The road was simply a mass of broken bodies, which were not even moved as the wagons were brought up, jolting over them! The Skipper sat his mount and surveyed the scene with a gleam of fanaticism in his eyes. He was splashed with blood, as indeed most of us were. It was one of the bloodiest corners I have ever encountered, and it stank!

Suddenly Captain Dupont began to laugh, and pointed with his saber at one of our comrades. He was a Swiss boy, named Dorneg, and he was engaged in an occupation at that moment which showed clearly that he must have gone *cafard,* for he was treading around the wounded Druses, a bidon in one hand and a Mauser in the other. His procedure was to raise the head of a wounded man, give him a drink from the bottle he carried, and then shoot him between the eyes. He was performing this office in such a cool, calm and unconcerned manner that, knowing the lad as I did, I felt sure he could not be aware of his senseless bestiality, his inhuman cruelty.

I dared not interfere, since the Skipper was only too obviously enjoying the performance. He let him go on, roaring with laughter as the half-wit went from one bandit to another, giving a sip, then a shot, a sip and a shot, until quite a number of the men and officers gathered to watch him and enjoy the sport. It was nothing, of course, when compared with the torture that wounded legionnaires had often been subjected to by the Druses; and yet it seemed so utterly senseless.

I was suddenly drawn away from this watching group by my chum, McCann, who rode up with his face in a ghastly mess. He had had a nasty slash. But it was not that so much which made me gasp at the sight of him. It was a look of something like horror in his eyes.

"Kim on!" he croaked. "Over here, in the gulley. Gosh, Barrington! You'll get the shock of your bastard life. Here it is!
Get down! Look!"

I dismounted and went over to the dead body he indicated, one of a heap.

"God in heaven! It's . . . it's Hoffman!"

"Surest thing you know, boy! It's Hoffman! Our old pal, Hoffman! And this goldarned crowd that we've just sliced up is the big army he started to drill a couple o' months ago. Can you beat it? His head's still on, but it's a miracle! Somebody plugged him as well. I wonder who. Gosh! What a bloody game this is!" "And we might have been with him," I said, staring hard at McCann.

"Yeah! Sure we might. Didn't he talk about his lil' girl, and her old man who had pots of dough. Remember him showing us the good crisp notes. Gosh, boy, an' here he is, stone dead from lumps of lead. Suffering Mike! Isn't it enough to make a guy walk right off into the desert and never come back? "

"And we'll do it, Mac. I'm through with this show, and I don't give a damn what happens. I'm going to slip, the first chance that comes along. You on? " "Betcha sweet life, boy! Shake!"

We did shake, a solemn vow, over the body of Hoffman. We took Hoffman's hand, too, before we finally covered up his face and left him. We dare not do more than that, dare not attempt to bury him, or we should have drawn the attention of others, many of the Squadron being still on the prowl for dead men's loot.

That was the beginning of the end of life in the French Foreign Legion for McCann and me. We had discussed the possibilities of escape thousands of times, in all sorts of places, but we meant it this time. We would quit whatever the cost. We had come to the end of our endurance. We were not built on lines as tough as our Skipper. Tough enough we were, and had proved ourselves. But we had limits.

What a victory for France! I wonder how many such victories she has had through the Legion of the Lost? It was all in life, Hoffman had said, that night, when we drank to our next meeting! Friends one day, enemies the next! All in life! But, merciful powers! What a life!

We fell in, a singing, jostling, laughing crowd who had won an unexpected victory that, actually, had been wholesale slaughter. There was no holding those jubilant men. The Commandant made no attempt to check them, as he would in other circumstances. Let them sing, let them throw their coarse jests to the prisoners, let them playfully prod the captured women helplessly shackled in their trucks, let them use their whips as they pleased! Did they not deserve their reward, these bearded men as easily humored as prattling children, who stank of gore and wine, whose eyes leered and whose lips fouled noisily over the women in the cart, savoring delights to come?

"Listen, Barrington. We'll get a chance to clean up after this. D'ja get me? That'll be our chance?"

I was riding some distance behind the Skipper, and McCann was alongside me. The Squadron was moving at a walk march, for our Commandant was taking things easy after this good day's work.

"We'll take some clobber round to the fountain in the Temple courtyard as soon as we're free, and we can discuss the details over the washing."

"You got it, boy. But say! What'n hell we going to do with the loot?"

I sat up with a jolt. Good Lord! Our loot was going to be one big proposition.

"If we're found in the desert with all that stuff in our packs, we should be murdered on sight."

"Dangerous stuff, boy. Damned dangerous. I guess we've got to do something about it. What say?"

"I should think we might take a little, of the smaller stuff maybe. It will be useful when we want to turn it into cash later on."

"Yeah? And what about all the rest? Gee, Barrington, there must be pretty nearly twelve thousand pounds' worth. I guess I know something about jewels and antiques, seeing it's my old Governor's trade. Boy! Think of my old boy, away there in Detroit, selling stuff at hundreds of dollars a time, and we gotta give the blamed loot away!"

"Give nothing! You big stiff! We'll bury it."

"Bury it! What'n hell for? That don't do no good."

"Would it do us any good to give it away? Wouldn't the boys suddenly get suspicious if we started handing out loot to them for no reason at all? We don't want to go round bidding good-bye to the Squadron, you sap!"

"Waal, I guess there's something in that."

"Besides, if we bury it properly, we'll always know where it is. You never know, you know! Perhaps, in a few years' time . . ."

"Haw, can that stuff! If you're thinking to bury stuff now and come back for it in a few years' time-waal, boy, you've got another guess coming!"

"Don't be a damned ass, McCann! I tell you we can bury this loot in one of those vaults under the Temple, and if nobody sees us it will be safe there for a hundred years, if you like."

"Yeah? Gosh, Barrington, I get you! I reckon the Druses will never use the Temple of the Golden Calf any more, now it's been polluted by us infidels, an' the French won't touch at all because of the burial ground there. Gee! I can see that place standing for centuries. Why, it's the slickest hiding place for loot in the whole damned Syrian country!"

"Of course it is! And we'll bury the stuff there when we go down with the washing."

"Betcha life! Look out!"

"Marche an trote/" bawled the Skipper.

We were moving up to the walls of Sueida by this time, a very proud Commandant at our head, a bunch of prisoners in our midst, and with us, too, the first women to enter the city since our occupation! No wonder the Legion cavalry rode in as if they owned the place! In a few minutes we should dismiss, since the Squadron must be given a chance to clean the grime and gore from their uniforms, and probably there would be no more duties that day.

It was as we had anticipated. In a few minutes we had handed over the prisoners, watered and stabled the horses, and McCann and I were making our way towards the Temple of the Golden Calf with bundles of washing.

The Temple is worth passing mention, for it was here where the Golden Calf had rested. What had become of it, when it was removed, or by whom, none seemed to know. It may have been taken away when the Druses evacuated the city. That, at all events, was the general belief. But I saw the huge marble plinth on which it had been poised. This still remained intact, and about the walls were mosaic effigies of the Golden Calf, together with a variety of inscriptions in Arabic. The Temple itself occupied only one corner of a huge courtyard, a kind of quadrangle, round which a balcony ran, supporting what were once the state apartments of the Sultan-el-At rash. These rooms were fashioned and decorated in a manner that would have delighted the soul of any artist. The interior walls were all picked out in colorful and pictorial mosaics, much after the manner of that which I had seen in Nashalla's house in Homs. One could imagine them beautifully furnished and plentifully decked with the rich tapestries and carpets of the East. But all these things had gone before we arrived on the scene. We could but stand within those noble interiors and try to visualize what they must have been like when occupied by the Eastern rulers and their suites.

But to the washing, which, with the necessary bath and delousing operations, we carried out at the fountain, that stood in the center of the great square courtyard. A base and ignoble use to make of such a fine piece of masonry as was that fountain, but needs must, and beggars cannot be choosers.

There was only one snag about that bit of washing and planning. We were presently joined by two German friends from the Squadron, Bess and Rehnke, who proceeded to divest themselves of their lousy shirts and tuck into the washing.

"I say, you chums'" commented Bess, who was a man of forty, and who had learned his English facing the Tommies in Flanders. "I say, you two chums, what do you think of clearing out and trying to get to Palestine?"

I stared at McCann. McCann also stood at gaze. Bess was a bit of a comedian, but he certainly didn't know he was.

"My chum here, Rehnke," jerking his head towards his pal while he scrubbed vigorously at his shirt, "he thinks we make good getaway from here because we are more South. Make Palestine easier, *hein?*" "Gottenhimmel," growled Rehnke, "we make him very easy. Sueida is good spot to get away. Senegalese make double guard to-night. They are not good soldiers. Bah! We have little trouble!"

"Awright!" groaned McCann, "but for the love of Mike, don't start arguing the toss about it 1"

"Yes," I snapped, feeling just as sore, "why the hell don't you get up on the balcony there, and shout good-bye to the boys before you go . . . *chums!*"

But those two hard-headed Germans who had butted in on our beautiful plans were too phlegmatic to appreciate our sarcasm, much less the reason for it. If they had made up their minds to desert this very evening the sun would be gone in another hour we should be unwise to desert as well, without warning them of the fact, for that might mean a terrible bungling of the long-anticipated event, and we dared not take any risks. Much the same kind of thought I could see, was passing through McCann's mind. I felt a kick on the ankle.

"Guess we'll have to take 'em with us," he muttered.

"Sure. Safest thing to do."

"Awright! Let's go! Now, listen, you bull-nosed Jerries, no hokey-pokey business. D'ja get me?"

"Mein Gott! You tink we play ta fool?"

"Only once, boy! Only once!"

"Listen, chums," pacified Bess, "we understand. All right. We make the plan for to-night. We go together, *hein?*"

"Got any money, Jerry?"

"We have little. We save a long time. No wine. No woman. Make little loot and make little money. We two chums all right. You? "

"Oh, yes, we're all right," I said. "We've been saving up, too, I don't think."

"Why did you join the Legion, *hein?"*

"Haw! Come off it, come off it! Don't start that all over again!" "Well, Jerry, why did you join?" I asked, seeing he had a story to tell.

"I tell you," He said solemnly, "I tell you."

"I know! I know!" shouted Mac. "You shot the other fellow's missis!"

"No, no! I shoot nobody's missis."

"Nope! This is a new one, eh? Awright. Spill it."

Bess, looking very solemn and very wise, blew out his cheeks, his eyes blazing with the thoughts behind them.

"They tell us we have five francs a day. ..."

"Five francs!"

"Yes, five francs! But wait. I say, very well, I go. But in Marseilles I meet German brother who is going home. He has one arm shot off. He tell me all about the Legion. Then I say, I will not go, *hein?* But they make me sleep.

I sign. Hacht! I come!"

"Hey! What'n hell's all this? D'ja mean they kidnapped you? "

That, in effect, was exactly what Jerry did mean. He and his friends had been drugged and kidnapped. We had come upon similar cases before. His was typical of a system that is in operation on the European Continent today. As I write these words, a report reaches me of two French agents being caught in the very act of carrying off a young German in a motorcar, whom they had rendered unconscious with a drugged cigarette.

It seemed that Bess, Rehnke and three other Germans, all hailing from Luxemburg, had been fired by the glowing tales of life in the French Foreign Legion. There had been promises of big pay (five francs a day!), and bigger prospects. Like other poor fools, they had sucked it all in, as a babe sucks milk the wonderful life of travel and adventure. They would very soon achieve the rank of officers, with plenty of money and wine and women, and endless opportunities for getting rich quickly.

Their fares were paid to Marseilles, and there they were taken to the Fort St. Jean, which is the depot for all Colonial troops. It was there where they met the German waiting for his discharge, he having lost an arm in Morocco, and he proceeded to give them a few home truths about the Legion of the Damned. Bess and his friends decided to go back home, for this was not the sort of life they had been looking forward to so rosily.

When they were paraded they refused to sign for enlistment, refused to fight for France with *Honneur et Fidelite.* Bess, who had appointed himself the leader, said they had been promised five francs a day, and now they had learned the truth from their compatriot, they had decided to go back home. Where upon half a dozen guards rushed in upon them and drove the unfortunate fellows into the underground dungeons at the point of the bayonet. They were left in the dark cells for forty-eight hours without either food or water.

Until they signed the enlistment forms they were free men. They were not criminals or bandits, that they should be thrown into prison cells. By that act, the military authorities at Fort St. Jean committed a grave breach of the law, and an even more serious offence in regard to international law. At Luxemburg those men were promised a square deal, instead of which they were flung into solitary confinement and went through hell. And this pernicious system obtains to-day. Press-gangs are working for France. Some French official must authorize them and reward them.

I know what it means to be confined in French dungeons. Bess and his friends were dragged out of their cells more dead than alive, rat-bitten, terrified, faint with hunger and thirst, for they had not received anything on arrival. They drank greedily of the coffee that was given them, and in that drugged and bemused condition they were given papers to sign and made soldiers of the Legion.

Bess told us all this, in his vivid, uncertain English, as we washed our shirts at that ancient fountain in the courtyard of the Temple of the Golden Calf. And as he talked, growing ever more passionate and guttural, our determination to make a bid for freedom that very night grew and strengthened. He and his friend, Rehnke, were welcome to join us. They were men after our own kidney. Good! Before morning we should be many miles away, across the plains, heading for Palestine!

We made our plans by that fountain shall I ever forget it! arranged the time of our meeting under cover of darkness, decided on what to wear and the kit to be carried. We were as excited as children at a party, though we contrived not to show it. Water was the main thing. We must collect some bidons, for we should need to carry at least three per man. As for arms and ammunition, it was impossible for them to draw theirs, but I, as officer's servant and bodyguard, always carried a rifle and Mauser pistol. We should be able to get along with those. They'd be useful in case of trouble. We had the advantage of the great majority of deserters in that respect . . .

As soon as darkness fell McCann and I collected our loot together, and made our way to the Temple. There were Senegalese guards on the outer walls of the courtyard, but they did not trouble me, as I often had to pass through the quadrangle with messages between General Andre and Captain Dupont. We had marked down a particular vault under the Temple for our hiding place. And there we silently buried the stuff, among the bones of Eastern nobles who had gone before.

There was nothing more to do now, save wait for the supreme moment! Midnight—and freedom!

CHAPTER XVII: ESCAPE!

Can you imagine what we felt like that night? We had got down to our bunks like the rest of the Squadron, but beneath our blankets we were fully dressed, with the rifle and the Mauser, the bidons filled with water, the mess tins full of rations, dates jammed tight in equipment, all waiting for the moment when we should creep out of this once mighty fortress, now practically in ruins, waiting for the magic hour of midnight.

I don't think I have ever suffered such nerves or been so excited in my life. This was my first, and last, attempt at escape! How many times had I discussed the possibility during those years, during that hell-on-earth period? How many hundreds of men had I known who made the attempt and failed? There was a palpable tenseness about those waiting hours. In spite of my fierce determination not to dwell on the problem, a whole procession of other such attempts would parade through my mind. And the inevitable question clamored for answer Should we do it? Should we succeed where so many had failed? Should we ever make Palestine? Should we ever see the first outpost of the British Empire with the Union Jack flying?

There came a sudden, swift tug at my blanket. A crouching form moved along the sleeping men. Now for the chance to practice that cat-footed tread of the Arabian a slinking crawl and no noise. I joined McCann and the two Germans outside. Mac, for all his cumbersome length, and the impeding rifle, moved like a ghost. Nobody spoke. The night was deathly quiet. It seemed that the slightest sound would give us away in that heavy stillness.

The night was pitch black, but we knew every inch of the road that led to the courtyard of the Temple, for we must cross this before we could reach the broken outer walls of the city, thence down the slope to the plains and freedom. There were double sentries at the courtyard, and double sentries again on the city walls. But, as we knew, they were all Senegalese, imitation soldiers, conscripted against their will, who spent their nights of guard either in drowsing or in shivering fear.

We came to the broken wall of the courtyard, and started to climb over. In places it was not more than three or four feet high. It was in fact still in process of rebuilding. It had been agreed that I should lead, from the start, and onwards. I led over that broken wall with a vengeance. I slipped, went over the top, and took half a ton of stone and dirt and dust with me! God! What a hellish clatter I made! It sounded to our strained ears like a score of French 75's all pounding at once. We all lay perfectly still, hardly daring to breathe. A little distance away we heard the startled, drowsy comments of the Senegalese. They put the noise down to prowling jackals, and we breathed again. Close to my ear as we lay on the ground, McCann bit off a dozen words:

"You blasted son of a!" Pause.

"You flaming1"

He was boiling with rage, and restraining himself with the utmost difficulty. My infernal slip might have cost us our lives. Had the Legion been on guard that night we should never have got away. A dozen shots would have been poured into the center of that clattering of stones, and no questions asked.

McCann crept about us, quietly clearing stones and rubbish before we got to our feet again. We crawled round the courtyard and then took the outer wall safely. We began a slow and careful descent of the hillside. On the plain immediately below us were more guards, Spahis Moroccan this time, patrolling the camp where the cattle were herded, and where the transport wagons and the airplanes rested. We made a wide detour of these and got past them without mishap. After what seemed like hours of back aching crouch, we stood upright and began to make swifter headway towards the open country. Once we had cleared the camps, McCann gave full vent to his heated, pent-up blasphemy. "'Of all the bastard sons of"

"Oh, shut up! We're not out of the wood yet!"

"Nope! An' we never should have been if we'd left it to you, Barrington! You"

"Don't make noise," muttered Bess, uneasily. "We all good chums together now."

We trudged on in silence. I went first with the Mauser, next came McCann with the rifle, and behind him the two Germans loaded with bidons. Mac might curse and blaspheme, but my heart sang as we proceeded under the thick cover of darkness, across the sandy wastes towards freedom. We were fit and fresh and full of hope, and it was arranged that we should march all night until dawn, then find shelter from the heat, the prowling bandits and the Partisans. We calculated that if we kept on in a south-westerly direction we were bound to make the Palestine Border.

Never shall I forget that night, the forced march, the sudden alarms, false and true, the arguments as to direction, the passing hamlets, where dogs would bark, the sudden snuffle-snuffle of the jackal packs that attempted to get in our way. We had to restrain an insane desire to yell and sing. Once we sighted a band of horsemen. They topped a ridge and came on towards us. We ducked behind a sand-drift. They passed us so closely that we might almost have touched them. What a terrible ten minutes that was! Suppose they had halted decided to camp on top of us for the night! They didn't. They passed on. We got to our feet feeling intensely relieved.

We were fully aware of what we were doing, fully conscious of the penalty of failure. We marched along knowing that every man's hand was against us. There was a price upon our heads. A hundred francs each four hundred francs for the capture, to say nothing of the amusement we should provide in that little matter of torture. And yet, that night, we were amazingly happy. We were tasting freedom. We believed we should get through. The risks and the hazards but added spice to the adventure.

It should be remembered that we had had years of training in the most adventurous clan in the world, the Legion. Adventuring had become second nature. We had taken our lives in our hands so often that we ceased to appreciate the fact. Now that we were adventuring in the finest cause the world knows freedom! we did so with a devil-may-care spirit ingrained by the conditions of the past few years.

It must have been that blackest hour before dawn when we met the Ghost.

“Down, everybody!” I gasped, and we flopped on our bellies in the sand.

Out of the desert came a shadowy form, a ghostly form all in white, advancing towards us slowly, meditatively but inevitably it seemed. Nearer and nearer he drew, until he was within a few feet of us. By this time I could see his face quite plainly above the snow-white burnous, his long beard, hook nose, strange eyes of the ascetic. He was wrinkled with age, but wonderfully upright. A Holy Man indeed! He appeared a trifle awe-inspiring at that weird hour. There were no weapons visible, but in his hand he carried a large, wooden staff. He stopped, gazed down upon us, and said:

"Why hide, my friends? I know you are there. You are even now thinking of shooting me."

"Waal, if this don't beat the band, I'm a son of a bitch! Why’n hell don't you shoot, Barrington?"

“Shoot yourself, you big stiff! You’ve got a gun! My arm’s paralyzed!”

“No shoot! No shoot!” whispered Bess.

The old man stood at gaze, regarding us with a strange but philosophic calm.

"Say, Father Christmas, what’s the big idea?"

"I am, my friends, Father of the Desert. For unnumbered years I have lived alone, watching my people wage a fruitless war against the infidels. You, my friends, belong to a great nation, but your wandering spirits have led you into a strange land to fight against my people, to fight not for the glory of your own countries but for that of another. Truly, you are strange people. Now you have grown tired of war and the shedding of blood. But come! It grows late. I have many miles to travel. Will you not give me a drop of your water and a portion of your bread?"

We gave him a water bottle and some biscuits. He sat in the sand, his head bent low in prayer, then began to eat and drink.

“Gosh! Can you beat it? Here we are, sitting in the middle of the desert entertaining a . . . a . . . waal, a bit of the Old Testament! Shucks! Ain’t we supposed to be making a dash for it? But I guess Father Christmas could do with a smoke. Here, have one of these, Dad!"

"No, my son. It is forbidden!"

The old man rose leisurely to his feet. His every movement was slow, measured, dignified. We rose, too, mechanically, rather taken aback by this totally unexpected visitor and his ineffable air of peace in that region of bloodshed.

"1 must be on my way. May Allah guide you, my friends. You will pass through many dangers as yet, and suffer much pain. But it is as Allah wills.”

We gazed after him, treading softly through the sand, his head bowed as if he were deep in prayer. We were too astonished to do anything but stare after the white shadow.

"Waal, Barrington, what do you make of that? Either we have gone plum *cafard,* or that guy hypnotized us!” "There are more things in heaven and earth ”

“Haw, shurrup! ”

"He said,” moaned Rehnke, "that we should go through many dangers and suffer much pain.”

"Now, now, Jerry! Nix on that stuff! You’re taking old Father Christmas too literally—d’ja get me? He meant before we die, not before we get to Palestine!” I wondered.

"Kim on! Let’s go!”

“Yes! At the double! ”

We started off at a run, our heavy boots slithering through the sand. But there was some comfort in that jog-trot. It helped us to make up for lost time, and, though none of us would have admitted this, it also helped to rid us of the sense of unreality that the old man had somehow infused. We kept it up for a couple of miles or so, but it was difficult going, for we had to hold our kit and guns and bottles close to our side in case we made too much noise, and that impeded our movements. We fell into a rapid march, now nearing the end of our first night of travel. Tomorrow night we ought to be within sight of the range of hills beyond which we should find freedom.

"Down, everybody!"

I dropped so suddenly that the others coming up behind fell on top of me.

"What in hell" N

"Look! Can't you see that light!"

Dimly from the distance came a faint murmur of sound and low on the ground was a glow of light. As we lay still the voices came down to us on the soft night winds, and peering through the haze we could presently discern movements about that dull glow.

"Partisans! They always give 'emselves away with their camp fires. Oh, boy! Wouldn't they be happy to know we were here!"

"Shut up, you fool! Sound travels!"

"Mein Gott! If they capture us!" Bess made a noise like a man vomiting.

"Shut your trap, Jerry. They will hear us! "

"What will we do?" moaned Rehnke. "We cannot *go now* . . . past them."

"If you blubbering don't dry up, I'll brain the pair of you!" Silence

for a space.

Now there was more disturbance of that distant glow.

"Gosh!" breathed McCann, "They've heard something."

"We must beat it," I said. "Come on! Let's go! Let's make a wide detour. Now! Sharp right angle and keeping moving, for God's sake!"

We changed our direction, moving at right angles to that dull glow ahead of us. We had to make some noise, however cautious our movements, for you cannot crunch sand without creating sound. I pushed on, the others trailing behind, and out of the comer of my eye I watched that glowing camp fire.

"If we come up against anything now, Mac, we'll use the knives, understand? The sound of shots would be fatal."

"Shoor, boy. I get you."

Suddenly that distant camp fire flared up as if it had been fed with petrol, which probably it had. Instinctively we dropped flat in the sand. Higher and higher blazed that flame of fire, lighting up the ground in an ever-spreading circle. We could now see the figures moving about quite plainly. It was obvious that they had heard us, but not surprising. An Arab with his ear to the sand can detect sounds a mile away.

In those tense moments we lay and perspired and prayed. There was just a chance they would take the wrong direction. After all, they had only heard us, they had not *seen* us. They were hardly likely to see us now. We were too far beyond the circle of light. I had made up my mind I would not be caught alive not by Partisans! I had seen too much of their handiwork. If the worst came to the worst, I would turn the Mauser on myself. Better dead. . . .

Lying alongside me, Rehnke panted like a man who had been running hard. His breath came and went thickly, so that we could all hear. Bess was making tiny sounds with his mouth, as if he were sucking in his breath in the fashion of a sufferer from asthma. McCann chewed dry Ups, and I tried to take an interest in all these things, because it was a period of waiting, just waiting.

The head-hunters were building up the fire with logs now. Some of their number had strayed beyond the further rim of light. "Time to move," I whispered. "They'll hardly hear us now . . . making too much noise themselves with the fire . . . and they can't see us.

Besides, it must be getting near dawn. That'll come up like a shot . . . and be the end of us! Come on! Let's go!''

We moved off at a run, going blindly, madly, as if indeed we were pursued by the Partisans. We kept it up for the better part of an hour. Then we saw the first streak of dawn. By this time we had left the camp fire far behind, had lost sight of it when we dipped into the shelter of a gulley. Here we came upon scattered boulders, a bit of friendly scrub, and harder ground. Of a sudden the sky lit up in bold, brassy colors, almost terrifying in their swift splashing of the heavens and the promise of intense heat to come.

And now for a rest! We had planned to march only at night and sleep through the heat of the day. Quickly we moved a number of stones over to a patch of scrub and built a shelter. It was barely big enough for the four of us, but we dared not be too ambitious, or we should become too conspicuous.

McCann and I kept a look-out through the crevices of our little enclosure, while the two Germans slept. Having slipped the Partisans, we were feeling thoroughly optimistic again. We had fed and drunk our portion of water, and were permitting ourselves a cigarette.

"I wonder who's cleaning the Skipper's kit this morning?"

"Gee, boy, I'd love to see him!"

"Or hear him!"

"Yeah. Won't that be pretty!"

"What will you do when we get to Palestine, Mac?"

"Oh, boy! What'll I do! I'll make for South America. There's bound to be a revolution down there. Anyhow, I could start one and make a pile of dollars!" "Haven't you had enough?"

"Barrington, you blighted limejuicer! D'ja think you could get on now without one? Boy, you'll just be sore for a scrap after a month's rest! "

I shook my head, enjoying the cigarette and the delightful interlude of peace. I could have wished this escape carried out differently. Try as I would, I could not drive Nashalla from my mind. Would I ever see her again? What was she thinking of me by this time? Where was she? In the hills behind Sueida? And all that loot now resting peacefully on the bed of a river that flowed through Homs? Lord! What a world! All that loot there, and more loot under a vault in the Temple of the Golden Calf. If and when I got back . . . well, should I find that cottage in sleepy Cornwall and dream about her and the loot?

"Couldn't do without a scrap now, after being in the Legion, boy."

"Haven't you any ambition?"

"Suffering Mike! Ambition! Shoor—to get blotto as often as possible, take a nice lil' cup o' tea occasionally, and enjoy life, boy. What say? "

"I always knew you were crazy."

At noon we roused Bess and Rehnke, had another meal, another careful measure of water, and settled ourselves for sleep while the two Germans took guard. They awoke us in the middle of the afternoon and were roundly cursed for doing so. A bunch of pariah dogs were shuffling and snuffling round our heap of stones, and Rehnke wanted to fire a shot and frighten them off. He was sternly forbidden. We did not want a shot echoing up and down that gully for any stray bands to hear.

A night and a day gone! We awoke to find the Germans grousing and bleating. They were blaming McCann and me for having led them into this wilderness, accusing us of not knowing anything of the direction. Apparently they had- been commiserating with each other while Mac and I slept, but we soon put an end to that.

"You can go back, or keep your mouths shut and follow us!"

I held the Mauser within an inch of the face of Bess, and McCann pushed the barrel of his gun into Rehnke.

"What say?"

We plunged into the night. I had a strong suspicion that we were drifting. Supposing those damned Jerrys were right! We certainly had altered our direction after coming upon the camp of the Partisans. I remembered things I had read about men lost in the desert, or in the bush, how they tramped round and round in circles for hundreds of miles, until they eventually went mad!

"Gosh! What's that?"

I swung round at the sound of McCann's voice. Something leapt through the air, knocking Rehnke to the ground. He let out an awful yell.

"Get up, you son of a bitch! It's only a pariah dog! "

Mac began to beat the animal with the rifle-butt, the sounds of padding feet and bloodcurdling snarls telling us that a hungry pack was closing in on us. This was the same pack that had snuffled hungrily round our shelter during the afternoon. They had waited for us! In another moment they were upon us. We were mixed up in a confusion of snapping, slavering, yelping mad dogs, and the weirdest fight I have ever encountered. Their green eyes flashed up as they sprang. I fired rapidly, bringing three or four to their knees, while McCann clubbed for all he was worth, and the two Germans struck with their knives.

"Kim on! We gotta run for it!"

Several made an attempt to follow us, until I turned and potted two or three. Then the beasts smelt blood, and fell on their wounded. "Hey, Barrington," panted Mac, "who said dog doesn't eat dog? Look at 'em, for the love of Mike! "

"They must have been hellish hungry to attack us."

"Betcha life."

"My shoulder," cried Rehnke, as we ran, "it is torn! One of those beasts got me. It is poison . . . death!"

After a while we halted and took a look at the German's shoulder. It wasn't much of a scratch, but as he said, it might be poisonous. We used a lot of precious water getting him clean, then set off again. All night long he grumbled and groused, complaining of his aching shoulder and the weight of his kit, until, thoroughly fed with his bleating, we had to relieve him of the bidons and rations he carried.

"Down, everybody!"

With my heart thumping furiously at these perpetual alarms, I dropped suddenly behind some scrub, and the others followed suit. I had heard the faint neigh of a horse, and though I could not see anything, I was taking no chances. Presently a party of horsemen came within sight, moving slowly, too slowly for us, as if the mounts moved cautiously for fear of awakening their riders. We crouched, waiting, guns ready. Was this another bunch of Partisans? Had they heard our shots when we fought off those damnable beasts?

But they came so near that we could almost touch their saddles, but it was obvious they were not looking for us, not looking for anyone, in fact. The horsemen were Druses, and across the saddle of each was a wounded man. Evidently they had had a brush with a French Patrol. We heaved a sigh of relief when they had passed on.

"Gosh, Barrington, another shave as close as that, and I'll die of heart failure!"

But the remainder of that night was uneventful. Dawn found us bivouacked under a bunch of rocks, McCann and I taking first turn-in while the other two watched. I was roused at noon by a punch in the ribs from McCann.

"Wake up, Barrington! *These lousy swine have drunk all the water!"*

I spent some seconds trying to take that in. It seemed an idiotic thing to say. *All* the water! Why, there must have been three full bidons when we went to sleep.

"Who's drunk all the water?" I asked stupidly.

"These two bastard Jerrys."

"Good God! You're crazy! "

I grabbed at the bidons, turning them over one by one. It was true enough. Every drop of water had gone. We had started out with seven full bidons, and had been on the trek two nights and a day and a half. Seven bidons! All gone! It should be remembered that, except for about half a bottle that was used on Rchnke's scratched shoulder, the water had been strictly rationed and was used for drinking only. There had been no possibility of a wash so far. We had not come upon any water in our travels, and God knows when we should.

I looked at Mac, then at the guilty Germans. I could guess what had happened. They had taken a drink or two while we slept. And that was a fatal thing to do in the heat of the morning. They had found they could not stop, had gone on and on until all the bidons had been emptied.

"Waal! What are we going to do with the swine! "

" Shoot them!'" I said, and lifted my Mauser to do it.

Mac pushed at my gun.

"We can't do that, boy!"

"Can't we, by God!"

"Nope! Cut it out."

"That's the law of the Legion, McCann; and you know it. They've stolen our water. We don't know when we shall get any more."

"We're not in the Legion now."

"Don't be so damned sure! I propose to give them a chance. Let them get on their feet, give 'em a running start and we shoot!"

"We can't do it, boy, we can't do that."

Rehnke let out a yell and began to jibber in German, his mouth foaming and his eyes bulging. McCann gave him a blow which quietened him. Bess, certainly the better of a doubtful pair, said not a word, but watched me from beneath his shaggy brows. "You know what it means, Mac? We've got to start on the march now, through this frizzling heat, and look for water. We have to keep going now, night and day. Two thirsty men travel better than four. That's horse sense, and you can't get away from it! What's come over you, you damned sap."

"Gosh, Barrington, d'ja think I'm not as crazy as you about the bloody situation these imbeciles have made?"

"Better let me shoot."

"Nope! Cut that stuff out. Kim on! Let's go! We'll strike water before sundown and fill up."

We set off, pushing the two Germans before us. We could not trust them anymore, so we walked behind them with our guns at the ready. And a weary afternoon it was, with the sun beating pitilessly down through sand-laden winds that seemed to have been forced from some gigantic furnace. Five solid hours we tramped through that blistering heat, without a halt, having seen nothing that promised water, not even a storybook mirage. Sullenly, silently, we kept it up, throats parched, tongues swelling. We breasted a hill, and decided to rest at the top. Rehnke was blubbering in a light-headed sort of way.

From where we lay there was a good view of the surrounding country. We could see a tiny hamlet in the distance. McCann closed his eyes for a short spell and I kept watch. The other two could please themselves. They were out of our calculations now. I stared through the heat haze at the faint outline of that village. It occurred to me that we should find water there but we must wait for darkness.

God in heaven! Was I dreaming? Surely that was an Arab leaving the village, ascending the slope diagonally. A solitary Arab. I was certain of him as he came nearer. He was carrying something on his back and walked with the aid of a staff. I dug McCann into wakefulness.

"Look! An Arab over there! Keep your eyes skinned, Mac, I'm going after him! He's almost sure to have a water bag!"

"Shucks! I'm coming, too!"

"Stay where you are, you fool! We don't want these Jerrys wandering off. They'd give themselves up first chance, and where the hell would we be then? You wouldn't let me shoot 'em, now we've got to carry them on our backs wherever we go. Hang on to 'em, now."

And with that I jumped to my feet and went after the Arab. I had not gone far when the fellow saw me approaching. He stopped, wondering who I might be, and waited for me. As I drew near I saw that he was dressed for travel. There was a pack and a water bag on his back. I did not hesitate for a second but went straight up to him, my knife concealed in my hand. He was giving me the usual "Greetings, brother," when I lunged. My onslaught was so sudden and unexpected that he had no time to guard. He sank to his knees in the sand, the blood dribbling from his neck. I felt no compunction then, nor do I feel any now. Many of his compatriots had done the same with comrades of mine, done it less mercifully, for he was dead before I had cut the pack and bag from his back. I made my way back to McCann.

"Did'ja finish him?"

"Of course. We can't have anybody running about to tell tales."

We filled four of the bidons from the bag, fed on the little beef that was left, and helped ourselves to the dates and bread from the Arab's pack. Among other things I had lifted was a lovely jeweled knife, a small bag of money, and some tobacco. All of which, as Mac said, would come in useful.

"The old Arab's water is as hot as hell, and probably fever ridden" commented Mac, "but, by God, it's like nectar to me!"

"Heavens! Yes!"

"Right! Kim on! Let's go!"

We trod down the remainder of that afternoon more lightly, giving the village we had seen a wide berth. Presently we emerged out on to a huge expanse of open desert with hardly a ripple in the flatness and certainly no sign of vegetation.

"Hey! What about this?"

We stopped and surveyed the scene. We were travelling in the glare of the sun, and if anything hostile came our way we should have difficulty in hiding, unless we buried ourselves in the sand. We decided to take the risk and try to make the hills we could see in the distance.

"Barrington, those hills mark the Border, I'm sure of it!"

"I wonder."

"Kim on! Let's make for them, anyway."

We trudged on, the Germans still in front of us, Rehnke keeping up an incessant muttering and mumbling, Bess sullen and silent.

"Oh, boy! We are making the Border with every step we take, and believe me, when we get the other side I'm going to swim in good wine for a week! Any guy who stops me waal, it will be too bad!"

And it was too bad! We went on for another hour. Those hills came nearer and nearer, and suddenly, over the top came a band of horsemen. We halted in our tracks, spellbound. We hadn't a ghost of a chance. There was nowhere to run, nowhere to hide. It was too late to get down into the sand. Those men had seen us, and were coming on, hell-for-leather. Rehnke sank to the ground moaning. McCann booted him to his feet again. I thought of the Arab. Had he been found? Such news would spread quickly. "Greetings, brother!" called the chief of the bandits, as they rode up.

"Greetings!" I said.

"Where are you going?"

"Palestine."

"Ah, escaped legionnaires, eh?"

If there is one thing an Arab loves, it is a gun. I tried to bargain with this chief. We were vastly outnumbered there were about thirty-live of them and something had to be done. I offered him the rifle and pistol if he would guide us safely across the Border. He wanted the arms first, and when I refused he attempted to pull his gun. But I was quicker on the draw than he. I fired and he toppled from his saddle. In a second we were surrounded. McCann and I were firing right and left. Not for nothing had we been trained as crack shots by Captain Dupont! Rehnke and Bess woke up and started to use their knives on the animals' flanks. The result was the whole band was thrown into confusion, and we were able to fight our way through them. Then we began to run, firing at the horsemen as they came after us. We were desperate men, men who had endured much, who had travelled two days and two nights to get thus far, and driven men can do much. We kept up that running fight, bringing them down as they came up with us, until the sands looked like a battlefield in miniature. We ran on and on, with McCann and I blazing away at any horseman who came abreast of us. For five miles we kept up that running fight, and we should have got clear, too, if the sounds of our firing had not brought *another* band of horsemen to the scene.

That was the end. We were through. We could not possibly stand up to any more.

"Kim on, you sons of bitches!" roared McCann. "Here we are! But, by God! We've given you a run for your money!"

A Lieutenant of Gendarmerie, riding at the head of the party, was enough to tell us that all was over.

"Cochons! Deserters, eh?"

"You've said it, you pup!"

"You will *go no* further!"

He rapped out an order to his men. Some of them dismounted and came over to rope us, hand and neck. I don't know how long we were in reaching the crest of the nearest hill, dragged at the end of those torturous ropes. We just staggered along as best we could, keeping pace with the trotting horses, feeling more dead than alive. Presently we reached a mud fort, and without further ado the Lieutenant had us all stripped and flogged. After which we were flung into a room together and a guard was posted over us.

We begged for water, and were kicked unmercifully for our insolence. Presently a sergeant of the Gendarmerie came in, looked us over, and selected me. He suggested that he had something to show me, and if I would give my word of honor not to attempt an escape, he would take me out on to the flat roof of the fort to show me this something.

"But, you know I can't escape.''

He thereupon took me up to the roof of this mud fort, from which I might look upon the country beyond the ridge of hills. And I saw, not three hundred yards away, the first outpost of the British Empire, with the Union Jack flying in the breeze. It was the outpost of which we had dreamed so long, and for which we had at last made our great effort. I stared at the Flag, flying there so bravely. Then I broke down. I wept like a child. . . .

CHAPTER XVIII: SENTENCE OF DEATH

The following day found us in Derra, a French frontier outpost some two miles or so from Tel Hadad, where we had been so ignominiously captured. Derra is the last post held by the French on the border line between Syria and Palestine. It consists of a scattered village, a well-guarded railway station, and a lot of barbed-wire entanglements, to say nothing of the extremely well guarded fort. Here is an almost dry nullah, but it has a trickling stream of sorts, over which is a bridge. On the Syrian side of the bridge is a French soldier, and on the Palestine side a British soldier. So near and yet!

Four miserable, dejected and utterly fed legionnaires were taken before the Commandant Bureau at Derra. A Commandant in the French Army is usually a Major. This Major was typical of his breed. He had us lined up in the guardroom, and then let loose his slimy tongue. His venom was rich in epithets. The names he bestowed upon us were many and varied. We were thieves, dogs, swine, murderers, and several other things which cannot be mentioned in an account that I have striven to make readable.

That Major might have saved his breath. His curses had no effect upon us whatever. We were too sick and weary. Besides, we were thoroughly inured to tongue slashings. We had soldiered under Captain Dupont for years. So he ordered the inevitable flogging. When a commanding officer of the French Army is at a loss—he orders a flogging.

It was here where we learned that we had *murdered* twenty-four Arabs! In our fight with the Partisans, when we had fought for our lives or our sanity rather than be captured and tortured, when we had put up a running fight across five miles of desert, it seemed that, in our desperate defense, we had killed twenty-four of them.

"Gosh! Twenty-four of'em, boy! Twenty-four! Canyon beat it? "

Even in my utter wretchedness, I could not help feeling a sense of supreme elation. I had sworn for years to get as many as I could, for all that I had seen of their diabolical handiwork. And I had done it, or, at all events, a goodly share of it. A wave of pride swept over me. I could have laughed in the face of that silly, cursing Major. Twenty-four! Wonderful! Wonderful! Twenty four fewer to torture and maim my fellow legionnaires! No wonder these cursing Frenchmen were annoyed! I had never felt such pride before, nor have I since that day. *Twenty-four!*

And so we stripped our already lacerated backs, and were hung up for another flogging. My recollection is that it did not take many strokes to lash the bunch of us into insensibility. We were too far gone to stand up to much. I wakened later to find myself rolled up into a ball, like a cat, but hardly so sleek and comfortable, for I was indeed bloody and messy. My first feeling, apart from the agony of the salt wounds, was one of rage, a rage that rose and swelled to ungovernable limits. I began to yell at the top of my voice, bringing down all the curses of Allah on the lousy French. Incidentally, I yelled for a drop of water. A sergeant came along and kicked me as I lay, until I lost consciousness again.

Somebody once said that man is the most adaptable of all the animals. I believe him.

The next time I woke up McCann and the two Germans were already sitting up and taking notice. Mac had torn a piece out of his shirt for a wash clout and was trying to clean the dried blood from his back with spittle. The other two were content to sit and endure. Now I had my eyes open I could look round on an indescribably filthy room. There was no light or ventilation, except such as might penetrate through the bars of the grill on the door, and the place stank. That might be us, I thought. We had not washed for days, and other attempts at hygiene had been of the most primitive order.

Still, we were alive.

Apparently someone had been watching us through the grill, for when it was seen that we were all awake, the order was given to march us into the prison yard. We were there only a few minutes, but long enough to see a man, apparently an Arab, being tied to the flogging post, and to witness some of his punishment. When I had looked at the fellow his eyes had answered mine with something familiar in them. I mentioned this when we were back in the cell.

"Shoor. I've seen that guy somewhere, but I've no nonce to remember."

I lay on the bare floor and pondered on that familiar look for a while. I fell asleep, I don't know for how long, but I was wakened by the sound of scratching close to my ear. Rats, I thought. But it wasn't the scratching of vermin, at all. A faint voice came to me through the wall, low down, near the floor.

"Englishman! Englishman! "

"Hey, Mac, what do you make of this?" We both listened and it came again:

"Englishman! Englishman!"

"There's a guy in the cell next door. Careful, boy. It may be another of their traps."

"Hulloa! Who's saying Englishman?"

"It is me, Taledio, the Italian, you remember, who used to serve you drinks at the *Cafe Noir?"*

"Good Lord! Then it was you I saw in the yard this morning . . . being flogged?"

"That is right, my friend. I am flogged every morning."

"Every morning! What'n hell for?"

"I cannot tell you now. It is that I am making a hole under the wall now, for you cannot hear unless I shout. To-night I will finish it. Then we will talk."

We lay and listened to the scratching. Hour after hour it went on. The little light from the grill faded out. Night had descended. Still the scratching went on.

"I wonder what that fellow's in here for, Barrington? He's a civi. He's nothing to do with the Legion."

"Must be something serious, if he's being flogged every morning."

"Yeah. I guess so. He's working hard enough at that hole. Hope he makes it big enough to get us all out of this lousy tinpot of a jail."

"Much good that would do us. Why, with these blasted handcuffs and leg-irons, we wouldn't get as far as the wall, much less get over it."

"You've said it. The lousy frogs are taking no chances."

Mac was still cursing the frogs when I fell asleep, if one could call it such. The heat was intense. Vermin played hell with our torn backs. The water we had been given was hot and stank.

I awoke with a start, trying to penetrate that hot blackness. McCann was leaning over me.

"Hey! You blighted limejuicer! He's through. Feel."

I felt where McCann's hand guided me, somewhere near the foot of the wall where we lay. A stick of wood was prodding its way through, up from the floor of earth. We grabbed it and began to widen the hole from our end.

Presently we had a tunnel through which we could talk in low tones.

I said I would get through, Englishman. I dig this hole long time, yes?"

"You sure did. What are you in here for?"

"I tell you. Then, mebbe, you help me."

"Hey! You ten cent Dago! How the hell can we help you? We may be shot to-morrow or the next day!"

"I do not think so. But even if you are, it will pass the time."

"Say! You're a cheerful sort of guy! You smell like a funeral!"

"Shut up, Mac! Go on, Italy. Let's hear you."

"I know lot of rebel chiefs in Damascus. They give me messages to pass to friends who come to *Cafe Noir.*"

"Yes. But what made you come to Syria at all?"

"Ach! It is a long story. I have one girl. She no good. She go with 'nother fellow. I kill."

"We've heard that one, Tally. Cut it out! What about these rebel chiefs?"

"One day big chief come to me. We make lot of talk about revolution. I give him much sympathy, yes? Then he tell me I make plenty of money with guns and ammunition across the Border from Palestine."

"And did you?"

"But, of course! I make a lot of money. Then I am betrayed to French. They bring me here. I am flogged every morning because I will not give secret place of guns, I know where five hundred rifles are hidden this minute, but they will never know. I will never tell. If I do they will find them, and shoot me. If I do not tell, they will shoot me ..."

Suddenly there came a piercing shriek. Quickly we pushed the earth back into the hole and rolled over it. But not quick enough. The door was flung open and in came a couple of guards.

The next half hour we spent in trying to evade the flying whiptails. We knew that something similar was happening next door. Having grown tired, they left us. We lay still through the remainder of that night. But in the early morning, just as the light began to break through the grill, we heard the Italian again.

"Englishman!" he whispered desperately, "if you have money, bury it in the floor. There is an Arab interpreter. He is my friend. You trust him to get help. He need money to go to Damascus. Druses have cut the telephone ..."

We never heard that stumbling Italian voice again. The money we had was quite safe sewn into the lining of our tunics. But this was news indeed about the interpreter. McCann and I decided to trust the fellow, whoever he was, providing he came our way. It was taking a long chance, but it was better than doing nothing. Since the telephonic communications were cut, the Commandant at Derra would not trouble to get in touch with the authorities at Damascus. He would probably take the law into his own hands, and shoot us for deserters. If this interpreter fellow could get through on horseback to the British and American Consuls in Damascus, there was just a chance that they would intervene, that they would demand a proper trial in Damascus. If I

After we had been given black bread and water, and the guard had retired, Mac and I took turns at the grill. We could see nothing except a dark passage and the cell opposite, but there was the possibility that the interpreter might come along. Presently, McCann, who was staring through the grill, beckoned to me. I joined him. Someone was coming down the passage. There was the unmistakable tramp of military feet, and into view came four French soldiers with fixed bayonets. They passed our cell, but halted at the next. We waited. They returned, taking with them poor old Italiano, his leg-irons clanging and clattering.

"Going for his morning constitutional," commented Mac.

"Yes. Poor devil! I wonder how long they will keep it up."

But we were wrong. We realized we were wrong when we heard that single volley fired. We knew only too well what that meant. Had we not fired many such volleys? Italiano had gone home.

"Gosh, Barrington, another guy on the long trail! I guess he's better off that way. He sure had a nerve not to spill the beans to the lousy French, eh?"

"Yes, but I think he might have told us who were the H.Q. on the other side if the guards hadn't come in when they did. They lost their chance there, too. And we got a licking for nothing."

"Nothing? Not if this interpreter guy comes along."

"If! If ifs and buts were apples and nuts ..."

"Hey, Barrington, you limejuicer! What's biting you?"

"I'm not building on any interpreter."

"Haw! Go to hell! What's the good of moping around here, groaning like a wounded stone-crusher. You'd better join the Jerrys in the corner there. That's your camp!"

"Never mind, Mac. We'll be able to make faces at the firing squad."

We were saved any further argument by the approach of the guard. It was time for our ten minutes' exercise. We had not been back in the cell more than half an hour when the interpreter did come along, in the guise of a priest. He told us he was from Palestine. The news of our capture had travelled far and wide. It was something new for an American and a Britisher to find themselves in such a predicament. He was ready and willing to procure a mount and go off to Damascus at once. We gave him the money and hoped for the best.

On top of him came the Commandant with an armed guard I don't know if he was afraid of us attacking him carrying a sheaf of papers and looking mightily pleased with himself. He informed us that we should be shot in forty-eight hours.

"You can't do it!" I shouted. "You can shoot an Arab or a Druse without trial, but you can't shoot a French soldier I"

"You deserted your post. You will be shot for murder and desertion. You will be shot without trial."

"You can't shoot an Englishman."

"English! Tcha! You are legionnaires."

"We are American, English and German. Shoot us, you lousy frog, and you'll start something."

But all that Mac got for his pains was a slash across the jaw. The Commandant retired, and our curses followed him. For the remainder of that day we lay about, getting on each other's nerves. The two Germans blamed us for the plight we were in. It was our fault, they said, since we made them desert. They bleated and moaned until we grew tired of hearing them and had to use strong measures to quieten them.

"There's plenty of time for that fellow to get through to the Consuls," said McCann. "And the Commandant is going to have the shock of his life."

"Oh, yes, time enough to reach the Consuls; but have they time enough to get here? Remember the wires are cut. The only way to reach us is by using fast horses. I doubt whether they'll do it."

"You would doubt, you sap! Pull yourself together, Barrington. You're about as cheerful as a canary with tonsilitis. I tell you, when the American Consul gets going, boy, you won't be able to see him for dust!"

"Oh, if it comes to that, I don't suppose he'll be any quicker than the British Consul. . . ."

"Quicker! Suffering Mike, you're not comparing anything British with American? Why, boy, you people don't know what hustle is. Let me tell you something. America is a bigger and better country than ever your darned old British Isles knew how to be.'' "Bosh! America was a colony when Britain was an Empire." "Shoor it was! But you couldn't hold it. We licked you out of it, boy." "You may have done. But you couldn't find a language of your own."

"Waal, as to that, I guess we're doing our best!"

The silly argument went on hour after hour, and, looking back on it, I have often marveled at its sheer childishness, the way we stormed at each other about our respective countries, the inane and utterly futile points we raised, so that once or twice we very nearly came to blows. It was, of course, just an expression of the state of our nerves, though we could not realize that at the time, and we were terribly in earnest.

It came to an end when Mac began to sing the Star - Spangled Banner, and I tried to drown him with God Save the King. This roused Bess and Rehnke, who struck up with the Wacht am Rhein. We kept it up until the guards came hurrying down the passage and hammered on the cell door. They threatened to separate us if we did not stop. As we did not wish to be separated, we gave in. Then came the night and merciful sleep.

The next day a French priest came to see us, and asked if he could do anything to help us on our way. It was on the tip of my tongue to tell him he could go to when Mac said:

“Shoor you can do something for us, Padre. Send in a gallon of wine."

"My boy, you are going where you will need no wine. Will you not pray with me? "

Rehnke broke down and sobbed like a child, while his chum, Bess, stared sullenly at the priest. He prayed for them.

"Say, Barrington, tell this guy to bring in some wine. If we *are* going on the long trail, then let’s soak."

When the priest had finished with the Germans, I addressed him in a mixture of French and Arabic, and apologized for my language.

"My boy, I speak English as well as you do," he said.

"You big stiff, Padre, why didn’t you do it before?"

"Ah, you boys of England and America are very brave fellows, but, I fear, you are far too wild and headstrong. ..."

"Never mind that, Padre. What about some wine? You know we’ve only got another twenty-four hours to live."

"Very well. I will try."

He was as good as his word. He saw to it that the wine was brought in to us and not pinched by the guard. Then he left us to get as drunk as we liked. And we did. We spent the day in drinking, cursing the French, the Druses, the Arabs, the Riffs, and every other race we could think of. We sang songs far into the night. Once or twice the guards came and threatened to move us. We invited them in to do it. Apparently they decided they would leave us alone. The Padre’s visit had impressed them.

I know that McCann and I reviewed the many days we had spent together, the tough scrapes we had been through, and the numerous places we had visited both in Africa and Syria. We grew maudlinly sentimental about the various rendezvous, the drinking shops, and the nice cups of tea we had encountered here and there. For nearly five years we had been together, carrying our lives in our hands most of the time. We talked of the Skipper and sentimentally referred to him as a fine fellow, a great fighter, a wonderful sport, and expressed the wish that we might shake his hand before we faced the firing squad. All of which was as idiotic as it could be, but excusable in those terrible circumstances. For we knew, while we sang and cursed and argued and slobbered over each other, that we were fiercely striving not to think of that dawn which was to be our last.

We had by this time given up all hope of the Consuls intervening. Either the interpreter-priest was a thief and a rogue who cared naught for us and the fact that we were to be shot, or it may have been that he had been prevented from reaching Damascus. After all, it was a thousand to one chance. Supposing the messenger were honestly trying to help us, it meant that he would have to reach Damascus with all possible speed and without being intercepted by bands of brigands, or Partisans, or French Patrols. And even if he got through, could he impress the Consuls with the urgency of our case; would they be able to get through the formalities quickly enough to permit them to reach us in time?

We talked and sang and joked and argued all in a vain endeavor to drown these questions that were never far from the forefront of our minds, in spite of the wine and the spirit engendered by it. Then the dawn came. We saw the light through that hellish grill. God! Shall I ever forget that dawn!

And to-day I sometimes hear men make that infernal joke:

"You'll be shot at dawn!"

Why? What is there funny about being shot at dawn?

We heard the tramping of feet, and we got to our feet, well primed, thanks to that priest. He understood. He was a man as well as a priest. A thousand thanks, Padre, wherever you are! You were human! But so few of your compatriots of the French military forces are human.

The tramping of feet. Only a few minutes to go now, surely? We staggered to our feet, swayed, hung crazily to each other.

"Give us your fist, boy. Looks as if we'd gotta go through the hoop, eh? " "Cut it out, you big stiff . . . 'member that girl I pinched from you in the Pit?"

"So! You *did* pinch her! You son of a gun, Barrington! If we hadn't got this . . . 'pointment . . . I'd put my foot through your clock!"

"Why, you big swaggering Yank, you couldn't put your foot through putty!"

"No? Is that so, boy? Waal, I guess it doesn't matter much now! We'll soon be hearing the angels . . . asking . . . what sort of harp we prefer . . . ."

Tramping feet! They seemed to be ages reaching our cell door, while we waited there, swaying up against each other, talking all manner of nonsense, anything at all so long as we could make a noise, keep back the damnable silence that precedes thought.

Tramping feet! Would they never reach us?

They stopped. There was a clatter of rifles. An officer's voice:

"Gardez votes!"

McCann and I gazed at each other. Should we ever hear that familiar command to attention, again? There was a rattle of keys and bars. The door was kicked open. The Commandant stood there, at gaze, a look of incredible hate in his eyes. Even we, in our fuddled, bemused state, could see that.

McCann gave him a mock salute.

He rapped out a command. We fell in.

"Right! You lousy frog!" roared McCann, his face a few inches from the Commandant's. "Kim on! Let's go!"

And to the amazement of each one of us, and most of all to Mac himself, the Commandant completely ignored the insult.

It was then that I became aware of something unusual in the atmosphere. What was all this? Why had not the Commandant turned on my pal and beaten him? It was an extraordinary position. Nothing like that had ever happened in my experience of the French military authorities. McCann was so astonished at not being half-killed for his insolence that he could not say another word, but instead, marched, or rather staggered, along that passage and out into the glaring sunshine of the prison yard.

We blinked as we stepped drunkenly out into that blazing light.

"Arretez !"

We halted in a fashion, reeling up against each other.

We waited, eyes on the ground to protect them from the sun's fierce glare, for they were inclined to water and weaken after that night of debauchery. We all looked up at the sound of approaching hoofs. What was this? Riding leisurely across the yard towards us were two mounted men. They wore civilian drill and topees, and they were covered in dust from head to foot, while their mounts were bathed in sweat. Every drop of blood seemed to drain away from my head. I went suddenly cold. Who were these two men? Who could they be, in that sort of clobber, save the British and American Consuls?

McCann let out a yell and ran forward to meet them, and I staggered after him. We must have looked like men gone *cafard* to those two horsemen, in our filth and rags, and the bestial state to which we had reduced ourselves during the night, added to which was the fact that we were suddenly bereft of speech. "It's all right, men, we arrived some little time ago," said one of them.

For the first time in my life, I fainted.

CHAPTER XIX: EIGHT YEARS IN CHAINS

The result of the intervention of the Consuls was that we were sent up to Damascus for trial before the *Conseil de Guerre.*

The jail at Damascus is a huge building that will accommodate upwards of two thousand prisoners, and it has a great expanse of flat roof, where the prisoners take exercise. We found it in an extraordinary state. There were no single cells in general use but a series of rooms opening on to the corridors. All the doors were left open, for the simple reason that the rooms were filled to overflowing, and those who could not find room on the cell floors flowed into the corridors and slept there. But that was not all. It was used as both civilian and military jail, and it housed all manner of prisoners, Arabs and Druses, negroes and Europeans, white, black and brown cheek by jowl, a conglomeration of races and nations all clanking chains, singing, swearing, chattering, in a score of different tongues. And over this mass there hung a hot, fetid atmosphere. The general conditions were indescribably filthy, and the food and water was poisonous. That wholesale system of flogging was still proceeding, just as it was before McCann and I left the place for Sueida. The beatings went on every day and all day long. I have never seen the bullwhip worked so hard. If one were not very slick, it was the easiest thing in the world, in that crowd, to get in the way of somebody else's beating!

"Waal, boy, you and me have been in some tough breaks, but don't this beat the band?"

"I am beginning to wonder whether we hadn't better be shot, and have done with it."

"Why? What do we want to do that for just to give these frogs the satisfaction? Nope! The Consuls blew in fine! And wasn't the Commandant pleased. Oh, boy, what a picture. . . . And you, you big stiff . . . fainted just like a limey!"

"You fainted, too, only the wall held you up."

"I wonder what sentence they'll give us, Barrington."

"A lifer, I expect."

"Cheerful cuss!"

"I wouldn't be surprised if they sent us to the Zephyrs."

"Gosh, Barrington, if they do, I'll *go raving cafard!* I'll create such a rough house as even that mob has never dreamt of." "You'll have something on, McCann. You know what the Zephyrs are?"

McCann nodded, suddenly quiet and thoughtful at the idea of being sent to the Zephyrs. And so was I. We knew something about the Zephyrs. Had we not from time to time met men who had done a stretch there? The world knows a great deal nowadays about the Devil's Island group of prisons, but who ever heard of the Zephyrs? Yet, Devil's Island, terrible as it is, is a playground in comparison.

France is a thrifty nation. That is the sort of thing I have had thrown at me ever since I set foot on English soil after my escape. I agree. She is. She never wastes man-power, particularly so in the criminal classes created by her Foreign Legion, and by her own unsavory underworld.

Why, says France, should a man of military age be absolved from serving in her Army just because he has been convicted of some crime, while decent boys serve their country? So some genius in criminal correction originated the Zephyrs, primarily for criminals of military age, and called the body *Infanterie Legere d'Afrique,* or, more commonly, the *Bats d'Af,* And the bats are, before they have been long in that terrible crowd. They get the name of Zephyrs because they are light infantry. They are also lightheaded most of the time. A big number of them are men who have come from the terrors of French Guinea prisons. They are given the opportunity to work out their salvation in the hell-spots of the desert. A man who is too hot for even the L6gion to hold him also goes to the Zephyrs. The guards are the biggest and brawniest of negroes, who thoroughly enjoy their powers of authority. Spreadeagling and such practices are common. If a man can endure the Zephyrs and remain sane, he must be a miracle of human endurance.

We decided we would raise this problem when next we saw the Consuls. They visited us daily and showed undoubted concern for our plight. Mac and I will never forget their many kindnesses. Among other things, they brought in cigarettes, not many at a time, since they would only have been taken from us, but enough to make those hellish conditions a little more tolerable. When we smoked, the other prisoners would gather round. It was pitiful. They clutched and clawed at each other, each striving to be nearest when the fag-end was discarded.

Then came the great day of the trial. The Germans went into the court first, and I cannot say what happened. I do not know what sort of case they made; but I can guess at what they said concerning McCann and myself. We had a young French engineer officer to speak in our defense. He came to see us, and we told our story. He promised to do his best. We, however, decided he was useless. He was obviously scared. Better no counsel at all than one who was too afraid to state our case. We informed the British and American Consuls of our decision, and asked for their advice. They could do nothing in this respect, except send in a lawyer to see we had a square deal, provided the French would allow it. Finally we decided to defend ourselves.

The two Germans were tried together, and they retained the young French officer as their counsel. McCann and I were tried together. The President of the *Conseil de Guerre* was Colonel Brevot, sitting with eight other French officers. Behind the President and his officers stood eleven guards with fixed bayonets, and above their heads was an emblem of the Republican Justice surmounted with the French Tricolour.

As we entered the President and his officers rose. We came to a halt and saluted. The guards presented arms. These ceremonies over, I was called. The following charge was read:
"Legionnaire 1384, you are charged with deserting your post, with enticing others to desert, with being leader of the prisoners and with killing twenty-four Arabs in cold blood.'" I smiled.

"Mon President, why not charge me with murdering the whole French Army, while you are about it."

There was a titter. The President flushed, glared, and cried:

"This is no laughing matter. It is serious!''

"I know it is serious for me, mon Colonel. But I suggest it is a ridiculous blunder."

"I don't want your opinion of French justice. Have you anything to say as to why sentence of death or life imprisonment should not be passed upon you? "

"Yes, mon Colonel. I have fought for France in eleven battles. I have served France with the motto of the Legion, *Honneur et Fidelite.* I have been decorated four times for my services. I am satisfied I have done my duty. With regard to the charges, I deny them all. At the time I deserted, it had been officially announced that the enemy was conquered and that peace reigned in the country. I was not the leader of the prisoners, simply because there was no leader. We had all planned to desert. There can be no question of enticement. If we killed twenty-four Arabs, I am glad. I am glad because they were Partisans whom I fought in self-defense, just as I have fought the Druses and the Riffs in the defense of France. I am glad to know that I killed members of a clan who have tortured and maimed my comrades for no reason at all and some of those comrades have been your own fellow countrymen. If ever again I have the opportunity to fight and kill Partisans, I shall do so. I shall welcome the chance to kill Partisans on any and every possible occasion. . .

I realized as I said those things that I was prejudicing my case, but I did not care. I knew that the best we could hope for was a long term of imprisonment, and, that being so, I made up my mind to gain some little satisfaction by saying what I felt to that President of the *Conseil de Guerre* for I should never have such an opportunity again. The whole trial was a farce from start to finish, for, as I have indicated, there was no trial, I was merely asked why the sentence should not be passed upon me. The President's mind was made up before we entered the court. He had heard all about us long before we appeared. And he was certainly prejudiced against us. We hadn't a dog's chance, anyway.

And I ended: "*I* defy the court to pass life sentence or sentence of death upon me. You dare not sentence me to death! You dare not sentence me to life imprisonment! If you do, it will be the biggest blunder France has committed in years! I shall never be shot by your authorization. I shall never serve the term of imprisonment which you have already made up your minds to give me. . .

It was sheer braggadocio, of course. But I did not care. I had been through so much at the hands of the French, and never had I a chance to speak for myself, until that so-called trial. I enjoyed myself thoroughly, so much indeed, that I felt the opportunity made up for many of my beatings and injustices. No one can appreciate, without being in identical circumstances, just how much I enjoyed that occasion. It was one of the supreme moments of my life, and the more Colonel Brevot glared and gaped at my audacity, the more I enjoyed myself.

But if my defense in that court was just a species of madness, what, in heaven's name, was McCann's? He was charged with deserting in face of the enemy, with aiding and abetting Legionnaire 1384, and with assisting in the murder of twenty- four Arabs. He was asked what he had to say why sentence of death or sentence of life imprisonment should not be passed upon him.

He denied the charges categorically. His long, lean jaw went longer and leaner as he thrust at the President and told him he was there merely to call his bluff! He became extraordinarily heated. In common parlance, he went off the deep end, plunging in with a recklessness and abandon that was good to witness. He asked the President point blank whether he thought *he* could stand up to thirty-five Arabs in the middle of the desert, with only three men supporting him, and get away with it. He wanted to know what the President would feel like, if, after fighting for his life against such odds, *and winning,* he were to be charged with murder?

"If I am to be charged with the murder of Arabs, what son of a gun in the whole of the French forces in Syria is *not* to be charged? I've killed Arabs by the dozen. I've killed Druses by the hundred. I have killed Riffs by the ton. If I were paid to do it, and I had on the right sort of uniform, I should kill you. What'n hell's all this business about a mere twenty-four Arabs? This isn't a charge. It's a farce! What have we been doing these past five years if we haven't been killing Arabs of one sort or another in self-defense? What's the French Foreign Legion for, if not to kill a few Arabs? Isn't a legionnaire paid to kill? Isn't he a Hell Hound of France, always wallowing in this dirty game of killing? What's the big idea twenty-four Arabs! When did France begin to concern herself about the lives of Arabs? She pays men to slaughter them every day in the week, and you, you, you can sit there and charge me with assisting to murder twenty-four Arabs. I didn't assist. I killed them with my own hands, holding a good rifle issued by the French authorities. I join my pal in saying that if ever I get another chance to kill Partisan Arabs, I shall kill them, and enjoy killing them, mon President. They are lousy scum or the French wouldn't employ them and the world is well rid of them, and all such. What's biting you, mon President, is that you don't know the difference between killing and murder. You can't murder an Arab, or any other man, while he has a better chance to defend himself than you have. Whadda mean, cold blood? Where does the cold blood come in, when thirty-five of your paid torturers rode four of us down who were on foot, who had been trekking the country for three days, two of the four being unarmed? What's biting you lousy frogs is that you don't know what a square deal is! If you think you can sit in that chair and sentence me to a lifer, go ahead! I am an American. My pal is British. We are legionnaires no longer. I'll tell you what you can do, Mr. lousy President. In plain American- English, you can go to hell!"

Then there was what the newspaper reporter would call "consternation in court." Colonel Brevot jumped to his feet, his face the color of a beetroot, and roared: "Take him away! Take him away 1"

With very little ceremony, they bundled the pair of us out of the court.

"Well, McCann," I said, "that's torn it!"

"Haw, shucks! Why worry, anyhow? They'll murder us in our sleep one of these days! "

The guards stood about us, three deep, and stared. McCann turned his sarcasm on to them. He had not spent himself in his outburst before the President, not by any means. The guards, with their fixed bayonets, stood and blinked at him, saying not a word as he lashed them right and left. That, in effect, was how we spent the half-hour of waiting.

We were called back into the court. The guards presented arms. The President rose and addressed each one of us in turn:

"Legionnaire 1384, I sentence you to eight years' imprisonment to be served in chains. Legionnaire McCann, I sentence you to eight years' imprisonment to be served in chains. Legionnaire Bess, I sentence you to four years' to be served in chains. Legionnaire Rehnke, I sentence you to four years' to be served in chains."

And before we had time to recover we were hustled out of the court and thrown back into prison.

CHAPTER XX: MUTINY IN DAMASCUS JAIL

The jail in Damascus is the world's worst. It is Gehenna itself.

It is divided into many stinking, verminous rooms, and at the time I was there, it held nearly two thousand prisoners, who also stank and were verminous. It was a hotbed of disease and vice, naturally. In there, men were piled in heaps, thrown together in rooms that knew neither ventilation nor hygiene. The unhappy band of four would-be deserters shared a room with forty others, a room that measured some twelve feet by eight, was in almost total darkness, and terribly rat-infested. There was no bedding of any kind. We used the bare floor and part of the corridor beyond the open door. If that door had ever been closed we should have died of suffocation in a very short time.

The food consisted of one pound of Arab bread per day per man, at eleven a.m. one pint of soup that stank, at five p.m. another pint that stank, together with filthy pieces of meat. We did not eat in those rooms. That was hardly possible. We went on to the roof for our meals and for exercise. The roof was flat, high-walled like any other prison yard, and beautifully fresh after the fetid atmosphere of those dungeons below. We rose at five a.m. and went on the roof, where we stayed under armed guard until eleven. After the meal of soup we returned to our rooms until four in the afternoon. At five we had our last meal of the day, and by six we had retired for the night, not to see the light of day until five of the next morning. What happened during all those hours below ground could never be imagined, and much of what went on is beyond description. Despite the fact that all the men were chained ankle and wrist, many of them managed to quarrel and fight.

I had seen many fights during my time in the Legion, but never any like those in Damascus jail. Many a morning we wakened to find a corpse amongst us. During the night a swift silent knife had fallen once. How these weapons were smuggled in was a mystery. I suspected the guards. Sometimes a prisoner grew too troublesome, despite his continual beatings. And then he died.

One night four Arabs were thrown into our cell. They were Syrian Gendarmes who had been caught selling information to rebel chiefs. At all events, that was the charge against them. Whether it was a true charge or merely a trumped-up one it is impossible to say. There was so much intrigue, chicanery, bribery and corruption that it was useless trying to get at the truth of things, half the time. When these four Arabs arrived amongst us they came fresh from a regular beating-up for they were bleeding like stuck pigs, and they groaned and moaned throughout that first horrible night.

For the next few days they spent their time quarrelling with each other, accusing each other of having brought about this terrible predicament. This perpetual squabbling got to such a pitch that we suggested they should fight it out and have done with it. They agreed. But their idea of a fight was vastly different from ours.

From heavens knows where, they managed to procure a knife each. One night, after the guards had retired to their own end of the corridor, preparations were made for the big fight. It was to be a silent affair of knives. The four Arabs were stripped to the waist and had taken the sandals off their feet, so that there should be no noise. It was an equal fight, two on each side.

What fine physiques they had! Someone had made a candle torch of fat, and by this feeble glimmer they faced each other, four brown muscular torsos already gleaming with sweat. They had only a few feet in which to make combat. They rushed at each other, swiftly, silently. There was a scrape of steel against steel, the sharp, hissing intake of breath, a soft shuffle of bare feet on the floor, but not a sound from the watching prisoners who crowded the corridor and the doorway. In that ghastly, almost silent display, one could hear the slither of a knife as it ripped, the sharp grating against bone. Very soon, the blood was flowing freely. One brawny figure went down with scarcely a sound, the other bent over him. There was a gurgle. . . . The other two continued to fight, but more and more feebly as the loss of blood brought the inevitable exhaustion. God! What a slaughterhouse it was! What a sight! What a stench!

Suddenly there was a warning cry. The guards rushed on the scene. We all scuttled back to our places. There was a great deal of shouting and cursing. But no one knew anything about those four bleeding men on the floor. And nobody cared. All four were now beyond human aid. It only remained for a fatigue party to pick them up and carry them away. No doubt a wrong had been avenged, and Allah was satisfied. We heard no more about it.

Life in Damascus jail was never monotonous, one was either being flogged or watching others endure that punishment, for the beatings were invariably carried out in the sight of other prisoners. I have never seen such a whirlpool of blood and lust, and I hope I never shall again.

McCann and I spent a month there, and God alone knows how we survived. The pig-trough food became so abominable that there was constant talk among the prisoners of mutiny.

They wanted a leader and often looked towards the Britisher and the American since we had a sort of reputation in the place. But what could they do, shackled most of the time with a yard and a half of strong chain?

It is true that at meal times an officer walked round asking for complaints, but as he was always accompanied by a sergeant major, who menacingly twirled a bullwhip in his great, powerful hands, there were never any complaints. There was no doubt at all that the officers responsible for rationing the prisoners of Damascus jail at that time were making a great deal of money, for they never bought anything but offal and refuse to serve up to the men.

At first, Mac and I had complained, but we soon learned, through repeated floggings, that it was unwise to be impulsive about such matters. We reported the state of affairs to our visiting Consuls, and we had another flogging for that. By this time there was real bad blood between us and the two Germans, so much so that I found it necessary to knock both of them out. And I was flogged for that. It did not, however, deter McCann, who also received a flogging for his man-handling of Bess and Rehnke. The human back is broad. It will stand innumerable beatings. That was a true anthropological fact beaten into us in the jail at Damascus.

Such conditions engendered all manner of crazy ideas in the minds of the prisoners. There was a suggestion amongst us that we should set fire to the place, but it was rejected on the grounds that, since it was practically all stone, we should do more damage to ourselves than to the building. Another brainy proposition was that we should make a concerted attack upon the guards during the night, take their arms, and fight our way out of the prison. But the French were wiser than us. There was that yard and a half of chain from ankle to wrist, always impeding our movements, and the only time we were unshackled was during the so-called meals, which were taken on the roof.

"Most of these hobos are *cafard"* commented McCann, as we listened to their plottings and plannings.

"And we shall be in the same boat if we stay here much longer."

McCann gazed at me with that queer, knowing look on his India rubber face.

"Yeah," he said slowly, "I get you, boy. I guess we have made a few rough houses, and we've always pulled through. . . . Waal, what d'ja know?"

"I'm thinking about the prison yard on the roof. There is only one door leading up from the building to the yard. That's why it takes only four guards to watch a yard full of prisoners. There are only two ways of getting out of that yard, one is over the wall and break your neck a few hundred feet below, and the other is past the four guards at that solitary doorway."

"Gosh, boy, we can't escape that way. You're crazy! Even if we could overpower four guards, we should then have to dive into the building, make our way through it, and out to the street below. And what'n hell d'ja think all the guards inside the building would be doing playing shove ha'penny? "

"Wait a minute, you big stiff! Who's talking about escape? Your blinking American hustle is overstepping itself. I'm suggesting we get the crowd to mutiny against the rotten food. They are all ready for it, anyway. I suggest that when the soup is served at eleven, and our chains have been unfastened, we attack the four guards, pinch their guns, and hold them there until somebody comes to inquire about the row. We can then demand to see General Andr6, and refuse to put down the guns until he comes to see the sort of grub we have to eat."

"Shoor! I get you. Count me in. Kim on! Let's get going! "

But before we could get the scheme into working order McCann and I found ourselves in trouble again. We were awarded ten days' cells for attacking one of the guards. After a visit from the Consuls, when we had been assured that we should never do eight years in chains, we had returned to our cell and were seated on the floor, talking, when I got a nasty jab in the jaw from one of the guard's rifles. Why that assault was suddenly made upon me I cannot say, except that I had been pointed out to this new guard as the troublesome Britisher. Anyway, both Mac and I jumped to our feet and struck at the fellow's stomach with our manacled hands we could not lift them higher on account of the dragging chains. The result of that little escapade was that we had a taste of the French solitary confinement. We were taken to a black dungeon, a place entirely without light, chained with a yard of chain to a damp floor, and there we stayed for seven days. Our food, which consisted of bread and water, was pushed through a narrow aperture in the wall. Imagine the cramped position, the unending hours, never knowing which was night and which day, our only amusement the rats that played around us. I have described in an earlier chapter how I endured forty-eight hours alone in a black cell. But that was a holiday compared to seven days' cells in Damascus jail. The only thing that saved our reason during that black and hellish period, that period which was timeless while we endured it, was the fact that we were together. Hour after hour we sang, talked, spouted, recited anything to make a noise, to shut out that awful black fear that we were blind and would never see our fellows again. There was another fear, too, the fear that we should lose the use of our limbs. We would raise ourselves into a crouching position, which was all the bit of chain would allow, and tentatively manipulate one foot, one leg, and then the other. And when we found that, after constant efforts, we could still move the muscles, we simply shrieked with joy. Then would come a period of playing a game of pretense.

We imagined we were strong men, and would tug and strain at our chains, which were secured to an iron ring in the stone floor. Looking back on it now, I can afford to smile at those strong-man antics. Our pretense at giving a feat-of-strength display before an imaginary audience was too funny, especially McCann's oratorical address:

"Ladies and Gentlemen, I am now about to show you how it is possible, with the aid of the superhuman strength nature has endowed me, to break this stout chain, link by link. ..."

But it usually ended with "Haw, shucks! It can't be done!" And we would lie back on the floor, falling asleep from sheer exhaustion.

Imagine the wrecks we were when, at long last, we were released! When we tried to stand upright we simply flopped sideways, for our legs were doubled up, numb and lifeless, and seemed to trail from the body as we heaved and wriggled about. The guards, however, soon restored life to the tottering limbs. Their idea of massage was to beat us with a rifle-butt, and very soon another sort of ache and pain was penetrating to the senses.

We were pushed and shoved along those underground passages until we finally reached our old cell, and there thrown to the ground, to recover as best we might. But the mooted mutiny and our eagerly-awaited return had made us many friends. Those poor fools of prisoners, who could not move without a leader, welcomed us as long-lost generals. Some had saved little tit-bits of food for us, another had stumps of cigarettes and matches clenched between his dirty toes—we enjoyed the smokes notwithstanding! others were anxious to massage us, and all made it clear that they would stand by us in any project we liked to take in hand.

"Gosh!" laughed Mac, in spite of his aching misery, "it's just like coming home!"

And ever afterwards he referred to those seven days as "the time when we went to the movies at Damascus and it was very foggy."

The prisoners were all for starting the mutiny next morning, but we turned the suggestion down. It was essential that we should have a few days* rest, so that we might get back to something approaching normal before we went into a stunt like that. After all, no one could tell what would happen, what might be the result. It was conceivable that we should be a great deal worse off after such an affair than ever we were before. Bread and water was no diet on which to train for such a stand. So we continued to coach the men for another week. Then the great day came.

The time was set for eleven in the morning, when the soup arrived. We went on to the roof yard with the crowd, about 250 from our section of the prison. Of course, Mac and I were two of the four chosen to grab the rifles. We saw at once that this part of the task would not be very difficult, as the guards were young soldiers. We all managed to appear perfectly natural, and the guard did not suspect. Our chains were loosened and the soup was given out. I rushed at the guard, snatched his rifle, and thrust the fixed bayonet towards him. McCann did the same and so did the other two. In the scrimmage a shot was fired by accident. Then we knew there would be trouble! The sound aroused the whole prison guard.

We, however, meant to have justice or die securing it. Up they came towards that solitary entrance which led on to the roof. We stood shoulder to shoulder, holding the rifles in readiness, a snarling ugly crowd of prisoners backing us. The Sergeant-major came, that powerful Corsican brute who loved to feel the shaft of a bullwhip in his hands. He found he could go no further than the entrance. So he stood there, demanding to know what it was all about.

"We want decent military food and no more floggings."

He laughed, but it was cut off short by the menacing snarl of the crowd behind us, anyone of whom would gladly have strangled him.

"Oh, my fine fellows. Decent food, did you say? Put down those rifles, or I'll shoot your rotten bones."

"Yeah," murmured McCann quietly, "and you come a step nearer and I'll shoot you for a dog!"

That Sergeant-major knew we were in dead earnest. He dared not put his foot in the yard. We told him to bring the Commandant, as we intended to have a square deal. He argued. He even promised to give us better food, but we knew what his promises were worth, and told him so. Finally he disappeared.

The Commandant arrived.

"Men, you are very foolish to carry on in this way. It can only end in bloodshed. Put down those rifles and go back to your cells. I promise you all a free pardon."

"Not on your sweet life!'' cried McCann. "We surrender only to General Andre, or General Gamlin, whoever is now General-in-Chief of Damascus."

"I warn you," spluttered the Commandant, "I will bring machine guns here and mow you down like dogs!"

"Go ahead!" I shouted, "and see what it will get you!"

He promptly sent for the machine guns, and we thought for a moment that he intended to use them, but only for a moment. We soon realized that this was just a species of bluff. He gave us two minutes to drop the rifles, or he would turn the guns on us. "You cannot murder 250 men," I said, "and remember, more than fifty of them are French subjects. If you fire you will kill your own countrymen. What will France say to that?"

We had our way. General Gamlin arrived and asked who was the ringleader.

"There is no ringleader. All we ask is justice and decent food."

"Ah! It is the Englishman!"

"Yep, you big sap! And don't forget the American!" "Now, men, what do you want?"

"Clean food and no floggings. If you will examine the backs of these men, you will find the marks of the bullwhip on every one of them."

"Men, in the name of the Republic of France, I ask you to put down those arms. I promise that no harm shall come to any of you."

It sounded good to us and we did as we were bade. Then we turned to the crowd and told them to remove their shirts. I don't know what General Gamlin thought, but I swear he never saw such a sight in his life as those two hundred or so lacerated backs. McCann and I went round with him, examining the results of many beatings. He was a shade pale as he turned quickly from one back to another. The picture was ghastly enough to make any man wince. After all, these were not the men whom the authorities were trying to get confessions of revolt from. Our section at that time was filled mainly with military prisoners. There were French soldiers here who someday would return to France. What a story they might tell! Not only that there was this Britisher and his American friend, who had already made a lot of trouble, had caused Consuls to intervene, had somehow managed to evade the punishment of death for desertion and murder. . . .

It was a strange experience McCann and I walking round with the General-in-Chief of Damascus on an inspection of a couple of hundred backs, the Corsican Sergeant-major marching in the rear with hate in his ferrety eyes, scowling at us whenever he dared. He knew that he was for it. And he knew that we knew. If looks could only kill we should have been dead men, up there, on the roof yard, where the sun poured relentlessly down, displaying a thousand scars, the handiwork of this Sergeant-major and his brother N.C.O.'s.

We knew exactly what he was thinking. He was thinking of the court of inquiry that was bound to follow this amazing inspection.

The General turned to us, not the Sergeant-major, and ordered us to marshal the men and go back to our cells in an orderly manner. We did so. From that day until we left the prison our food was eatable, and floggings were considerably fewer.

Obviously they could not keep McCann and me there after such a disturbance. So that we were not surprised to learn of our banishment from Syria altogether. We were to proceed to Beirut, and there take ship for Marseilles our stretch of imprisonment was to be served in France!

CHAPTER XXI: BEIRUT AND BEYOND

A few days later we received the order to collect kit for the transfer to Beirut. We were chained together as we marched through the streets of Damascus to the station. Arabs and other inhabitants stopped to stare at us, some crowding round us and following a little way, their women spitting at us, until a thrust from one of the gendarmes drove them off. At the station we were locked in a compartment of the train, and the journey from Damascus to Beirut of fifty-five miles was uneventful, save that we stewed in a close, stifling heat, which must have been somewhere near no degrees.

A very pleasant surprise awaited us at Beirut. Our new prison was beautifully situated on a hill overlooking the sea. It was the cleanest prison I have ever seen, and I have seen the inside of twenty-two. The cells were spotless and spacious, about twenty feet by twelve, with plenty of fresh air coming through the wide barred windows. It was all too good to be true! Our reception was almost unbelievable, for the British and American Consuls had already sent us fresh fruit and cigarettes!

"Oh, boy! Take a load of this! Can you beat it! Good smokes, fresh fruit luscious and fit for the gods! good clean beds and no beatings up! Waal, you lousy limey, if this isn't heaven, I'm a polecat!"

"Heavens! I just can't take it in, Mac!"

"Nope! Waal, sit tight and watch me!"

And it was so. For three glorious days we ate bread that was clean and sweet, fresh vegetables and appetizing meat; and were allowed our ration of smokes. In those three days we managed to get ourselves cleaner than we had ever been during the preceding three years. As McCann said:

"Gosh! I wouldn't mind doing ten years in this crib!"

Of course, it was too good to last. Before we had time to recover from our astonishment at this humane treatment we were shipped down to the quayside and put aboard a dirty, stinking cargo boat bound for Marseilles. We were stowed away in a dirty hold, chained to the ship's side with about four yards of chain, the only light a thin shaft that penetrated through the hatch above. Our escort, French soldiers going home, left us there without food or water for twenty-four hours. Later we learned why. They were naturally overjoyed at the idea of returning to France, and had spent that time in a carousal with the Captain of the ship.

Meanwhile McCann and I stewed in the heated atmosphere below. We had made the acquaintance of rats before, but down in that hold they were better and bigger than any we had ever seen. And they were fairly tame. At all events, we managed to capture two, which we secured with pieces of twine and then set them to races, for which we wagered our cigarettes. We passed an hour or two that way, then grew tired of the game, and began to shout again. No one took any notice of us. Perhaps they could not hear our yells. Perhaps they were too busy with the hooch, as Mac would say.

In fact, our escort were so busy with their own affairs that they had not troubled to search us after coming on board. That gave us an idea. We found a place between the planks in which to hide our cigarettes and matches. We did that with a two-fold purpose. The smokes would be safe there, and we could take them out whenever we had the chance. Also, we husbanded the stock of matches very carefully, for we had conceived the scheme of setting fire to the ship when we judged we were somewhere within reach of Cyprus. Once the boat was fired, we should be unchained and put into a lifeboat and then would come our final effort to escape. In spite of what those Consuls had said, it really looked as if we were destined to do eight years in chains in some obscure French prison, and, rather than risk a landing on French soil, we decided to take this big chance.

Once we had made up our minds regarding this plan we thought of little else. At the end of our four yards of chain we could roam about the hold with a fair amount of freedom. There were bales of jute at one end, excellent material for making a fire! There, we decided, operations would begin. We should have to judge the nearness of Cyprus as best we could, but, as we figured it out, we had only to wait for the ship to get under way, and once she was out of harbor we must "feel" the same lapse of time as that between Damascus and Beirut, since the port was about the same distance from Cyprus as it was from Damascus. That was not very good seamanship, but the best we could do in the circumstances! If we fired the ship too soon, or too late, we should have to correct that when we took control of the lifeboat, as we meant to do.

Cyprus, from what we had heard of it, was an island that provided unlimited places for landing. What we should do when we reached the island we did not know. We never got so far in our plans of escape, the main thing always was to achieve the escape, and this seemed the best scheme yet! I am afraid we never gave a thought to the crew, the dire peril in which we must inevitably throw everybody, nor of the boat and its cargo. We had spent five years in conditions which had bred and fostered the instinct of every man for himself and the devil take the hindmost. After such a life, such planning and scheming came very natural to us. Escape! Escape!

Escape! We never lost sight of possibilities in this direction. It became a mania. For were we not facing eight years in chains?

The shaft of light disappeared, and we slept the sleep of the just and the hopeful. During the following day we even went so far as to move some of the bales, not out of the actual stacks, but sufficiently to make the job easier, when, anticipating another night of solitary confinement, we should set fire to the ship.

But it did not work out that way at all. Does it ever? That evening the guards who were supposed to be escorting us to France came down into the hold by way of the iron ladder from the hatch. Our first intimation was when the removal of a few of the hatch planks threw a big patch of light down to us. The sergeant of the escort, drunk as a warrior, came first, and it is a marvel he did not break his neck. Simultaneously with their coming we felt the unmistakable throb of the ship's engines and the shiver and sway of her as she left the quayside and got under way.

So the voyage had begun! I could not help feeling a thrill as I thought of the great scheme. I could have laughed! This was going to mean freedom, freedom at last!

The guards brought us a pail of soup and two loaves of bread. We ate ravenously while they stood about, armed and with fixed bayonets. When we had finished, McCann, wise in his hour, had the temerity to ask the sergeant why we had not been attended to before!

He got a boot in his belly for insolence. But as he remarked to me afterwards, he had had a feed first, so he was more or less protected. That night was a perfect farce so far as our plans were concerned, for the guards, instead of leaving us, went only as far as the hatch cover and sat there with their legs dangling, looking down on us, their rifles pointing downwards at us! Waiting for darkness was no use, either. When that came, an electric lamp with a wired shade was slung over the hatch, so that we could not move without being seen. There was nothing for it but to go to sleep. Perchance, next day we should get our opportunity.

Next morning, however, just as dawn was breaking, the sergeant and three seamen came down to us. We were told we should have to work our passage at the stokehold. If we attempted anything we should be shot. The Chief Engineer and his men were all armed, and they had the Captain's orders to shoot at the first sign of insubordination or mutiny. Our chains were unlocked and taken from us. We were given half-an-hour's exercise on deck, and we thought how glorious it was to taste the fresh sea air after being down in that rat-ridden, stinking hold for twenty-four hours.

Then we were kicked along to the fires down below, and there we slaved at the insatiable fire-boxes for a twelve-hour stretch, stopping only once for food, when we had more soup and bread. All the time we were under the eagle eyes of the armed guard, which was frequently changed in that hot and dry atmosphere. There appeared to be, besides the sergeant and a corporal, about four or five privates in the home-going party, all of the military gendarmerie. To look at them, one would hardly believe that these men had tortured old men and women in order to extort information from them. Now they were happy, laughing and chatting about their homes, their families, their lovely country. It is not difficult to imagine what we felt like, working there, half naked, shoveling coal into those hot, glaring, gaping fire-boxes. We raged, of course. We cursed and swore as we worked. If we straightened our backs for a second, somebody's .bullwhip would descend. . . .

One would have thought these home-going soldiers would show a more lenient attitude towards men such as McCann and me, but they were as brutal as any officer or N.C.O. of the Legion. No doubt they had heard all about us before taking over the Britisher and the American who had murdered twenty-four Arabs, who had been saved from death at the last moment by the intervention of their countries' Consuls, who had created a mutiny in Damascus jail, who had made a perfect nuisance of themselves everywhere they bad been, who were known to be very desperate characters! We were more determined than ever to fire the cursed ship. If we did not, if we allowed this boat to carry us and land us at Marseilles what earthly chance should we have, with our reputation? "Barrington! We've just gotta bum these lousy frogs out of it!" groaned McCann, as he labored with his shovel, the sweat pouring from him in streams, as indeed it poured from me.

"Yes. And we'll do it to-night."

At the end of those twelve hours, however, we were too utterly worn and weary to worry about anything. After being dragged along to the shower bath, we were chained up in the hold again, where we instantly fell asleep, the heavy, dream less sleep of the exhausted. And so it went on, day after day for a week, shoveling coal all day long, weary, exhausted sleep all night long. Even so, I doubt if we could have carried out our plans, for the guards never left us after that first twenty- four hours.

Actually, we were relieved to sight Marseilles. Not that we had given up hope of escaping. In our most depressing moments, we never did that. We looked towards France. Who could say what might happen there? After all, men *did,* escape from prisons, even though they were chained. If we had to do eight years in chains, we could not be kept in the shackles every moment of that time. No, we had not lost hope, even though the firing of the ship was a frost!

CHAPTER XXII: SHALL I GO BACK?

So we arrived at Marseilles, the refuse tip of the world, the city of contrasts, that is neither East nor West, but a hideous conglomeration of both. We move in, worming our way through liners, tramps, steamboats, full riggers, barges and every litter of the high seas. Chained, we stand on deck, under guard, awaiting our new escort, viewing the multifarious life of the amazing port longing, longing that we might be let loose for just one little minute!

The guard arrived and we were at once conveyed to the station, where we entrained for Albertville. We were fifteen hours in that train, travelling through very beautiful country, though we had no opportunities to admire it. We travelled in a convict carriage. Down the whole length of the carriage was a central corridor patrolled by warders, and facing this were tiny compartments only big enough to hold a man in a sitting position, and there we sat, ankle chained to wrist, forced to a crouching position, one arm left free to take the mug of food pushed through a hole in the door. We were taken out in turn and marched up and down the corridor for exercise. My turn came twice during that fifteen hours.

On arrival at Albertville, we were marched with other prisoners through the old-world village to the prison, which did not look at all uninviting, since it had once been a monastery and still preserved its appearance of peace and seclusion.

The warders took McCann and me before the prison Governor as soon as we got inside. Much to our surprise, he seemed a kindly sort of fellow we should not have been surprised to receive the usual kick in the stomach, so accustomed had we become to the treatment. Moreover, he spoke excellent English.

"Legionnaires Barrington and McCann, I have heard all about your case, but I hope while you are here you will conduct yourselves properly. I assure you, if your conduct is good, you will find little to complain of here. That is all. Take them away."

We started life in Albertville Prison with six days' cells, and separate cells at that. This solitary confinement, however, was child's play compared to what we had been accustomed to. Our food consisted of two pints of water, one pint of coffee, two pounds of bread, soup at eleven in the morning, and soup with potatoes at six-thirty in the evening. The eternal darkness was damnable. Nothing to do for six days and nights but think, and, my God, how one can think in the darkness of a cell!

One morning my cell door opened and in walked the M.O., accompanied by the Governor and two warders. I was blinded by the sudden glare of light, and pushed my fists in my eyes while the warders unfastened the chains.

"Sortezvous !"

I staggered out into the corridor, obeying a blind impulse to get away from that terrible blackness rather than obeying the command. The medical officer ran his eye over me, and asked how I felt. I told him I was all right, except that I wanted some fresh air and God's good daylight. I was astounded at this inspection, this concern as to my condition of health! Nothing like it had ever happened to me after solitary confinement.

I was conducted to a huge prison yard, where I was put to stone breaking. With me, as with every other prisoner there, a long chain was attached to the waist, allowing me thirty yards of freedom. I soon got accustomed to its presence, and settled down to a life of stone-breaking, day in day out, stone-breaking. Later I was put in a gang to work the nearby quarry.

We did ten hours there every day, and as long as one's back was bent there was no trouble, but if one attempted to straighten up for a breather, down came the whip. I worked quietly and steadily. I kept my mouth shut. After all, the food was eatable.

One had a bed of sorts. I had known infinitely worse times. Besides, I was only biding my time. I knew, as sure as the sun would rise tomorrow, that my chance would come. Meanwhile, I was a well-behaved military convict who still had time to serve in the French Foreign Legion though I had not the faintest intention of ever going back there. I carried on like that for two months. Then came the Red Letter Day.

There was a general holiday that day in France for the Anniversary of the Republic. As a special concession, the convicts were freed from their chains for a few hours, and allowed to mix freely in the prison yard. Naturally, I took the opportunity to look for McCann. It seemed years since I had seen him. He saw me first, however, and came forward at a run.

"Hey 1 Barrington, you son of a gun 1 Not dead yet!"

"McCann 1 You big stiff! How are you? "

"Listen, boy. I've got some news. My Consul is still working for our release. He thinks he can get this sentence quashed."

“Which means we’ll have to go back to the Squadron and finish our time in the Legion!”

"Listen, boy. We’ll see to that when we get away from this goldarned crib. What we’ve got to do now is keep good.

Get me? It's going to be easier for us if we have a good record here, see? Promise me, Barrington, you’ll keep your end of the bargain. The point is, these guys are going to throw a revolt. D’ja get me? But we’re keeping out of it.” [11] Sure we are! You can count on me, Mac.”

‘‘Right! Now there’s five hundred desperate men here. You wouldn’t think it, to look at ’em. But they are, boy. I know. As soon as the Governor and the guard come up to put ’em back in the chains, they’re going to rush them, get the guns, and shoot their way out.”

The men were free until four o’clock in the afternoon, and as the time drew near I begun to feel quite excited. McCann and I kept together. This was to be a meeting in which we should take no part. We were warned by some of the old lags of what was coming. They never suspected that we should not join in. Had we not a reputation for revolting against law and order? There was not a convict in the place who did not know our story, though how they had come by it was a mystery, since neither McCann nor I had ever whispered a word concerning ourselves. We were never in the habit of bragging of what we had done.

It was getting near four o’clock and there was a tense look on many a face by this time, though everyone was trying his utmost to appear normal. Suddenly the whistle blew. The Governor came into the yard accompanied by the guard. His appearance was the signal for one mad rush. The prisoners rushed upon the guard, and there was a terrible scrimmage inside two seconds. The men just massed themselves against the Governor and his guard, getting in each other's way in the most hopeless fashion. Mac and I stood back by the wall and watched. God! What a scramble it was! Some of the convicts had captured rifles and there was a close-quarter duel between them and the guards who were still armed. Then we noticed a group of five or six wrestling with the Governor in an attempt to get his gun.

"Gosh! Look at that, Barrington. Kim on! We stand in on this. The Governor has been decent enough with us. Let's go!"

We ran up to the group clawing at the Governor and began to hit out right and left. In the general melee we managed to knock out one or two, and as soon as the Governor had freed himself, his revolver spat again and again, until they began to drop all round him.

Suddenly, firing opened from the roof of the building behind us and we knew the troops had arrived. It was all over. The big chance for five hundred convicts had come and gone, for some of them there would never be another chance, for others no chance would ever be needed, for quite a number had gone hence.

The crowd, now cowed and dejected, were lashed into their cells, all except McCann and me. We were still standing by the Governor, Mac with a cut eye and a swollen mouth, while I had lost a few more teeth. We were conveyed to another part of the prison, right away from the general crowd, though our chains had been locked on to us again.

For the following three weeks we were segregated from the crowd, taking our exercise alone, doing odd jobs of work merely to keep us occupied. The Governor had reported our conduct. It appeared we had saved his life, and he intended that the fact should be known in the right quarter. We were treated more like guests than prisoners, save that we always wore our chains. There came a day when we were called into the Governor's office the most momentous day of my life, I think, for it proved to be my first step towards freedom.

"In return for coming to my aid at a very critical moment," said the Governor, ''I have recommended you both for a free pardon. You saved my life, and that is something I shall never forget. I cannot thank you as I would like. I wish it were in my power to release you from the French Foreign Legion, but I cannot do that. That is beyond my powers. But I have done all that could be done with regard to your eight years' imprisonment in chains. I have here the

authority from the President of the French Republic, giving me power to release you from prison. You are to go back to your unit. I hope you will prove good soldiers for France. I thank you, Legionnaires Barrington and McCann."

And with that he shook us by the hand, expressing the wish that we might meet again someday in better circumstances. I came out of that office not knowing whether I stood on my head or my heels. McCann and I uttered not a word as we returned to our cells. There was no need for speech. One look was enough. We knew that this was the end. We had been told that we should leave the prison separately, and proceed to Marseilles separately. I was to go that very day. Mac would follow a week later.

In short, we were free men once more, soldiers of France proceeding to their units and moving freely without the watchful eye of a guard. I have never seen McCann since that day, when, before returning to my cell, we silently and solemnly gripped hands, while a look pregnant with meaning passed between us. In effect, that look said:

"Take your chance! I'm going to take mine!"

A few hours later I was back in my legionnaire's uniform, had drawn some pay from the local garrison, was escorted to the station by a corporal, and put on the train for Marseilles. I travelled in a compartment of the train, just like any other passenger a free man, my heart singing with joy. I knew that this was the end. I could scarcely contain myself. But I must! I must keep my head, now! Dear God! If I made a slip now. . . . But there should be no question of slips.

I was a free man, a soldier travelling to Marseilles alone. Heavens! How could there be a slip now? I lay back luxuriously in that carriage, gazing at the landscape flying past the window, my mind a tumult of thoughts, trying to keep calm, trying to feel like a free man, a soldier of France proceeding alone to Marseilles, there to take ship and rejoin his unit in Syria.

As dear old Mac would say:

"Not on your sweet life, boy!"

When my train steamed into the station at Marseilles, a corporal would search the train for me, a corporal detailed to escort me to the Fort of St. Jean, where I should receive my orders for the ship. But the station at Marseilles is a very big one. There are crowds of people there always, a bustling throng, and among them many troops returning to their various units. That would be my great opportunity. Far better than trying to slip the train before I got there. Even a soldier, even a legionnaire, may lose himself in the crowd at Marseilles, for are not soldiers, even legionnaires, constantly passing through that dirty city? Slip? God in heaven, what slip could there be now?

Already I could picture in my mind's eye that corporal detailed to find Legionnaire Barrington, searching in vain for Legionnaire Barrington! I could see him returning to Fort St. Jean, believing that we had missed each other, expecting to find that I had already reported!

And when I did not report, the corporal would shrug his shoulders, knowing that I had slipped him in order to get drunk before returning to service! It was all so beautifully simple save for the fact that I was wearing the uniform of a legionnaire!

Well, that should be attended to. I had a few francs. I knew my Marseilles! I knew more than the "Can o' Beer," as the seamen say of the Cannebiere. I knew some of the crooked little back streets. One can do anything in the Vieux Port, even change a suit of clothes! Lord! I could not sit still during that long railway journey. Up and down, up and down, a restless soldier of France returning to his unit, just one leave-lorn traveler unwilling to return. Unwilling! Determined never to return! This was the end—one way or the other! Even if there was a slip, and I was carried on to the boat, I could dive off. But, no! Not that end! There must be no slip.

As the train steamed into the station I dropped from the footboard, my heart thumping with excitement. In a few seconds I was skirting the fringe of the crowd, a few more, and I was out on the street. Free! Free! Free! I never hesitated, but strode on and on, making for the quarter that would in a few minutes turn me into a civilian. I came to the place I sought, dived in, purchased my dungarees, and in a miraculously short space of time I stepped out again, wearing my dungarees, cap and scarf, a typical docker, loafer, water- rat, scrounger, what you will. I slouched along, my gait a suggestion of a roll, my stride a trifle longer, a honest-to good- ness pipe in my mouth. I was by no means out of the wood. And I knew it. But surely it needed but a little more acting now? There had been no slip so far. But if I were caught in these clothes, back I should go to the prison chains.

Caught! I laughed, feeling good. I felt I would fight ten thousand corporals rather than be taken now! All I wanted was a British ship, and I didn't care a damn where it was bound for, so long as it *was* British.

I found my way down to the docks, and strolled nonchalantly around. I saw the cargo boat I wanted, and, heaven be praised, the Blue Peter was flying! Good! So far! The problem was how to get aboard. I was not the only would-be stowaway watching that ship. I recognized others, and I hoped I did not look as suspicious as they. How to get aboard? There was a quartermaster standing at the head of the gangway. No use trying to get by him. He was too wily a bird. He had seen my sort many times before. It wanted a couple of hours to darkness yet, and even then the gangway would still be guarded—if the ship had not sailed!

Heavens! Something must be done quickly! The longer I stayed in Marseilles, the less chance I had of escaping. It was now or never! My heart stood still for a minute. I gazed, fascinated, towards the stem end of the ship. Two steel doors had been opened and a small gangway thrown up to them. Up this gangway went a little procession of dock laborers, carrying cases and sacks that clearly indicated a last-minute rush of victualling stores. I carefully attached myself to the party, picked up a case, shouldered it, and proceeded up the little gangway. I followed the man in front, who led me along an alleyway to the ship's stores, where I dumped the case, and turned about as if I were leaving the ship for more. I trotted along that alleyway a couple of feet behind my fellow dock laborer. Behind me was another. Heavens! What now? If I dived into that little alley I had passed when coming aboard, this fellow behind would get suspicious, perhaps talk. That little alley came nearer, nearer, the sweat was pouring from me, so that I tore off my scarf to mop my face, accidentally dropped it, marched on. There was a shout from behind. The fellow pointed to my scarf. I returned to retrieve it. He passed on! Lord! Could anything be more simple?

Turning into the side alleyway, I found it ran athwart the vessel and led into another on the side of the ship furthest from the quayside. Then I began to wander about, dodging round comers or racing back, whenever I heard footsteps. Finally, I came upon a steel door partly open. I stepped inside and carefully closed the door. I guessed I was in the ash recess.

I sat down on the steel floor and breathed a huge sigh of relief. I ought to be safe there for an hour or two. . . .

There I stayed until the ship was well out to sea, then I stepped boldly out and almost fell into the arms of the Chief Mate.

"Hello, froggie, what the hell are you doing here?"

"I'm no froggie. I'm an Englishman. What's more, I am an escaped legionnaire, and I appeal to you as an Englishman to help me!"

"Well, I'm damned! Is this a yarn you're telling me?"

"It is not! Look!"

I took out my papers and handed them to him. One glance was enough. He marched me off to the Captain. I found myself telling my story, the whole story, of my life and adventures in the French Foreign Legion. The Captain listened, tugged at his thick stubble of beard, and repeated "Aye, aye, aye," at regular intervals. It was impossible for me to guess whether he believed my story. He never gave the slightest indication, just tugged at his beard and repeated that broad "aye." When I had finished he turned to the Mate and said:

"Tak' him a awa'. He mun work."

I gathered that the Captain was Scotch and that he never wasted words. I certainly worked during that short voyage between Marseilles and Newcastle, but I never worked so happily in my life. It was a pure joy! My story was all over the ship in very little time, and the crew with whom I worked could not shower enough good things upon me. They insisted on my taking the proceeds of a "round-up," as they called it, when, after the happiest voyage of my life, I went ashore at Newcastle.

And only a few days ago I received a letter from McCann, inviting me for the umpteenth time to join him in South America. Apparently he has realized his ambition, if ambition it is, and apparently he will never rest until I go out to him.

But, I wonder! My ambition does not lie in that direction, really. I suppose it will sound like madness if I admit that my thoughts are on Syria, on that wonderful creature, Nashalla? And how shall a man control his thoughts? Direct them into other and more sane channels? I have tried that . . . and found so many disturbing cross currents.

Shall I go back and look for her?

I wonder!

END

Made in United States
North Haven, CT
07 November 2026

21928352R00129